Post-Harvest Physiology and Preservation of Forages

Related Society Publications

Forage Cell Wall Structure and Digestibility

Forage Quality, Evaluation, and Utilization

For more information on these titles, please contact the SSSA Headquarters Office; Attn: Marketing; 677 South Segoe Road; Madison, WI 53711-1086. Phone: (608) 273-8080, ext. 322. Fax: (608) 273-2021.

Post-Harvest Physiology and Preservation of Forages

Proceedings of a symposium sponsored by C-6 of the Crop Science Society of America. The papers were presented during the annual meetings in Minneapolis, MN, 1-6 Nov. 1992.

Co-Editors
Kenneth J. Moore
Michael A. Peterson

Managing Editor
David M. Kral

Managing Editor
Marian K. Viney

CSSA Special Publication 22

**American Society of Agronomy, Inc.
Crop Science Society of America, Inc.
Madison, Wisconsin, USA**

1995

Copyright © 1995 by the American Society of Agronomy, Inc.
Crop Science Society of America, Inc.
Soil Science Society of America, Inc.

ALL RIGHTS RESERVED UNDER THE U.S. COPYRIGHT
ACT OF 1976 (P.L. 94-553)

Any and all uses beyond the limitations of the "fair use" provision of the law require written permission from the publisher(s) and/or the author(s); not applicable to contributions prepared by officers or employees of the U.S. Government as part of their official duties.

American Society of Agronomy, Inc.
Crop Science Society of America, Inc.
Soil Science Society of America, Inc.
677 South Segoe Road, Madison, WI 53711 USA

Library of Congress Cataloging-in-Publication Data

Post-harvest physiology and preservation of forages : proceedings of a symposium sponsored by C-6 of the Crop Science Society of America / co-editors, Kenneth J. Moore, Michael A. Peterson ...[et al.].
    p. cm. — (CSSA special publication : 22)
  "The Papers were presented during the annual meetings in Minneapolis, MN, 1–6 Nov. 1992."
  Includes bibliographical references.
  ISBN 0-89118-539-9
    1. Forage plants—Postharvet physiology—Congresses. 2. Forage plants—Postharvest technology—Congresses. I. Moore, Kenneth J. II. Peterson, Michael A. III. Crop Science Society of America. Division C-6. IV. Series : CSSA special publication : no. 22.
SB188.P7   1995
633.2'086—dc20   95-6087
                                                                                CIP

Printed in the United States of America

CONTENTS

	Page
Foreword	vii
Preface	ix
Contributors	xi

1 Post-Harvest Physiological Changes in Forage Plants
    Lowell E. Moser 1

2 Microbiology of Stored Forages
    Craig A. Roberts 21

3 Field Curing of Forages
    C. Alan Rotz 39

4 Hay Preservation Effects on Yield and Quality
    Michael Collins 67

5 Legume and Grass Silage Preservation
    E.H. Jaster 91

Foreword

Production and preservation of high quality forages has long been a concern for producers. This publication is jointly sponsored by the Crop Science Society of America and American Society of Agronomy. It represents the most current knowledge about preservation of forage crop quality as it is influenced by post-harvest physiology and microbiology. Producers, agronomists, and crop scientists will find the information in this publication to be beneficial and useful, particularly as it relates to field curing of forages, and hay and silage preservation.

The co-editors of *Post-Harvest Physiology and Preservation of Forages*, K.J. Moore and M.A. Peterson, are well-recognized for their contributions to the current state of knowledge in this area of forage research. The authors, M. Collins, E.H. Jaster, L.E. Moser, C.A. Roberts, and C.A. Rotz are leading scientists and educators in their respective disciplines. Their background and expertise are well-suited to integrate the most current knowledge on the complex subject of forage crop preservation.

This publication will serve as a technical reference for teachers, researchers and producers, who are interested in forage quality and utilization. It will be helpful in identifying, understanding and managing factors associated with forage crop losses in quality and quantity from the time of harvest through the time of its use. The technical content of this reference should beneficial for years to come.

Robert (Bob) C. Shearman
President, CSSA

Preface

The preservation of forage crops is one of the most risk-intensive processes undertaken by farm managers. From the time that a forage crop is first cut until it is used as feed, it is subject to significant losses in quality and quantity. These losses are incurred through a complex set of biotic and abiotic processes that occur during harvesting and field operations, and later during storage and handling of the product. To minimize the risk associated with forage preservation it is important to understand these processes, how they interact with one another, and how their effects can be mitigated through various management practices.

This special publication is based on a symposium of the same title that was sponsored by the Crop Science Society of America and held during the 1992 annual meeting in Minneapolis, MN. The objective of the symposium was to integrate knowledge from several disciplines as it relates to the preservation of forage crops. This publication brings together for the first time in one document the current level of knowledge in this important area. It is intended to be useful to a broad audience ranging from forage producers to scientists working in the general area of forage quality and utilization.

The editors wish to express their gratitude to the symposium planning committee who developed the original outline for this publication and to the authors for their efforts in the preparation of the chapters.

Kenneth J. Moore
Iowa State University, Ames, Iowa

Michael A. Peterson
W-L Research, Inc., Evansville, Wisconsin

CONTRIBUTORS

Michael Collins Department of Agronomy, University of Kentucky, N-122 Agricultural Science Building—North, Lexington, KY 40546.

Edwin H. Jaster Department of Dairy Science, California Polytechnic University, San Luis Obispo, CA 94307.

Lowell E. Moser Agronomy Department, University of Nebraska, 352 Keim Hall, Lincoln, NE 68583.

Craig A. Roberts Agronomy Department, University of Missouri, 210 Waters Hall, Columbia, MO, 65211.

C. Alan Rotz USDA–ARS–USDFRC, Room 206, Agricultural Engineering Department, Michigan State University, East Lansing, MI 48824.

1 Post-Harvest Physiological Changes in Forage Plants

Lowell E. Moser

University of Nebraska
Lincoln, Nebraska

When forages are cut for hay or silage, physiological changes occur in the plants that result in a certain level of unavoidable nutrient losses. These post-harvest compositional changes occur in forages prior to removing the dry or wilted forage from the field and in the forage material that is in storage. With hay, the changes in storage under good conditions are generally minimal, but major compositional changes may occur when forage is ensiled. This chapter will discuss the plant physiological changes that occur from the time the forage is cut until the plant material is no longer living or moved to storage. Compositional changes that occur later are brought about primarily by the physical environment or microorganisms that will be addressed in later chapters.

PREHARVEST MORPHOLOGICAL AND PHYSIOLOGICAL STATUS

Forage standing in the field may vary greatly in composition and in physiological activity. The physiological activity occurs in the protoplasm or living portion of the plant (symplast). The nonliving portion of the plant (apoplast), such as the cell walls, once formed, have no intrinsic physiological activity. They may affect the physiology indirectly by modifying water removal and interacting with outside forces such as microorganisms. Forage leaves are much more metabolically active than stems and contain much of the protoplasm of the plant. Leaf blades consist of mostly thin-walled mesophyll cells in legumes and cool-season grasses protected by an epidermis with a waxy cuticle. Warm-season grasses have a higher proportion of vascular tissue and bundle sheath cells. Stomata may be located on both leaf blade surfaces, through which gas exchange can occur readily. Leaf sheaths and stems are not as metabolically active as the blades. Leaves of both legumes and grasses dry more quickly than the stems, hence, metabolism ceases more quickly in leaves. Stems vary in composition and are not harvested in grasses until stem elongation occurs. When immature, both grass and legume stems contain nonstructural carbohydrates and protein. As stems mature, the cell

Copyright © 1995 Crop Science Society of Agronomy and American Society of Agronomy, 677 S. Segoe Rd., Madison, WI 53711, USA. *Post-Harvest Physiology and Preservation of Forages.* CSSA Special Publication no. 22.

walls become highly lignified and nutrients become less available. Lower stems may store a considerable amount of nonstructural carbohydrate, but in most perennial grasses and legumes these storage areas are not harvested with the forage.

A forage crop standing in the field, possibly involving several species, is a combination of tissues, with varying cell soluble and cell wall contents. Leaf to stem ratio changes with plant maturity. For example, Albrecht et al. (1987) showed that the leaf to stem ratio of alfalfa (*Medicago sativa* L.) dropped from approximately 1.3 to 1.4 at the vegetative stage to ≈0.5 to 0.7 when the alfalfa was in the full bloom to early pod stages. The cell contents stayed at ≈800 g kg^{-1} in the leaf portion regardless of maturity, but the cell contents decreased from ≈650 g kg^{-1} in vegetative alfalfa stems to ≈450 g kg^{-1} for alfalfa stems at the early pod stage. In grasses, the transition from vegetative to reproductive tillers can change leaf content drastically. Averaged across 2 yr, perennial ryegrass (*Lolium perenne* L.) leaf blade content declined from 85 to 20% of the dry matter as the canopy went from early vegetative to the fully headed stage (Minson et al., 1960). Buxton (1990) observed cell solubles ranging from 350 to 400 g kg^{-1} in reproductive cool-season grass stems, while leaves of these grasses ranged from 450 to 550 g kg^{-1} cell solubles. Lower leaves in the canopy had ≈400 g kg^{-1} cell solubles. Protein content varies widely within and among forage species. Alfalfa may contain >300 g kg^{-1} crude protein when vegetative, but it drops to ≈150 g kg^{-1} at full bloom stage. Temperate grasses often contain between 100 to 200 g kg^{-1} crude protein, while 50 to 100 g kg^{-1} crude protein is more common in the tropical grasses. The greatest concentration of protein is in the protoplasm of the leaf cell. It is comprised of a relatively soluble component (many of the enzymes) and an insoluble component that may be particulate enzymes or protein associated with membrane structure. The crude protein (CP) fraction is often comprised of ≈75 to 85% true protein, with the remainder in other forms such as amino acids and amides. In many situations true protein is ≈10% less than reported CP and this may drop to as as much as 25% less in young rapidly growing material (Lyttleton, 1973).

Glucose and fructose are the main reducing sugars in forage and sucrose is the major nonreducing sugar. Fructose polymers (fructans) are present in temperate grasses and starch is the major available polymer in tropical grasses and legumes. These carbohydrates are readily metabolized by plants after cutting and represent material that is nearly 100% digestible to livestock. Smith (1971) reported that amylopectin is the major type of starch that is present in alfalfa leaflets. Amylopectin is a large branched starch molecule of ≈2000 to 220 000 glucose units. Amylose is a linear molecule that generally is present in smaller amounts. Amylose is a smaller molecule comprised of 50 to 2000 glucose residues. Fructans, comprised of fructose residues added to a sucrose molecule, are small polymers ranging from just a few fructose units to several hundred, depending on species. Herbage from various cool-season grasses contains 80 to 100 g kg^{-1} total nonstructural carbohydrates (TNC), while legume herbage contains ≈70 to 110 g kg^{-1} of TNC (Smith, 1973). Stems of legumes and grasses often contain higher levels of sugars than leaves. Cool-season grasses also have higher levels of fructans in the stems; however, warm-season grasses and le-

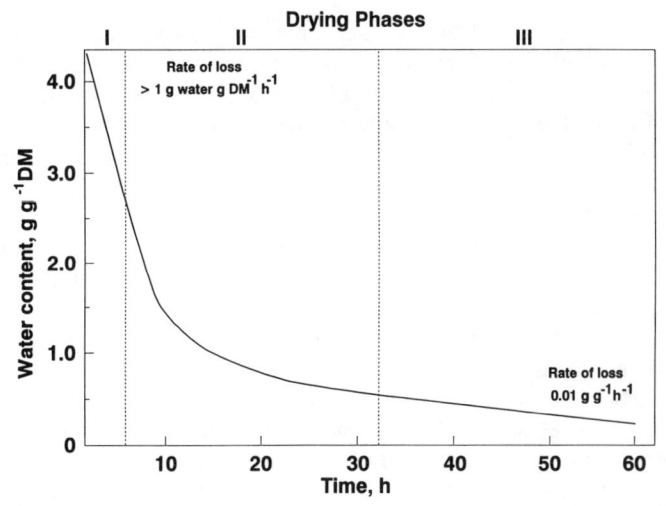

Fig. 1–1. Typical drying curve (thin layer, temperature 20°C, relative humidity 50%, air speed 1 m s^{-1}; Jones & Harris, 1979).

gumes often have higher concentrations of starch in the leaves than in the stems (Smith, 1973). Starch levels may be especially high in leaves late in the day or when cool nights occur. Holt and Hilst (1969) reported significant diurnal fluctuations in nonstructural carbohydrates. Lowest values were at 600 h and highest values were at 1800 h. Considering both the water soluble and 0.1 M H_2SO_4 hydrolyzable carbohydrates (starch), alfalfa increased ≈40 g kg^{-1} and cool-season grasses increased ≈50 g kg^{-1} in total nonstructural carbohydrates from 600 until 1800 h in July. Plant physiological and morphological status interact with harvest environment to bring about post-harvest physiological changes.

THE DRYING PROCESS

Three phases of field drying of cut forages are described by Macdonald and Clark (1987). These phases can be seen on the drying curve by Jones and Harris (1979; Fig. 1–1). The first phase involves rapid initial drying that occurs when the forage is high in moisture, the stomata remain open, and the vapor pressure deficit between the drying forage and the air is large. Initial water loss rate may be on the order of 1 g g^{-1} dry matter (DM) h^{-1} (Jones & Harris, 1979). Water evaporates rapidly from the leaf lamina of both grasses and legumes, drawing some stem water with it. Harris and Tullberg (1980) reported that detached leaves dried 1.5 times quicker than leaves on intact plants. When the osmotic pressure of the guard cells drops, stomata close and remaining water loss must occur through the epidermis and cuticle. Up to 70 to 80% of the water in a forage crop may remain after stomatal closure (Harris & Tullberg, 1980). Under good drying conditions Phase 1 is rather brief. The second drying phase lasts longer and involves cuticular evaporation of water. Leaf structure, cuticle characteristics, and plant structure affect the duration of Phase 2 (Harris & Tullberg, 1980).

Disturbing the cuticle on leaves or stems hastens the water loss. Plant metabolism continues and Phase 2 can be prolonged if the forage is dense, the relative humidity is high, or if there is poor air circulation. After the moisture falls below 45% (DM basis) the remaining water becomes increasingly difficult to remove (Nash, 1985) so that in the final drying phase, water is held more tightly in the plant material. Phase 3 is often extended by high relative humidity around the forage. Although plant metabolism has dropped to a low level in Phase 3, the forage is much more susceptible to damage from outside environmental factors such as shattering and rewetting. Phase 3 continues until the plant material is dry enough to be stored as hay.

The rate of water loss in grasses depends on tiller morphology as well as water content. Grass leaves dry 10 to 15 times faster than the stems, with as much as 30% of the stem water lost through the leaves (Murdock, 1980). Vegetative tillers with 80% leaf content dried in one-third of the time required to dry tillers with emerging heads that had 40% leaves (Jones, 1979). After head emergence, drying time is reduced due to lower water content and increased exposure of stems. Harris and Dhanoa (1984) ranked drying rates for headed Italian ryegrass (*L. multiflorum* Lam.) tillers as follows: leaf sheath > leaf lamina > whole tillers > exposed stems > inflorescences > enclosed stems. Harris and Tullberg (1980) provide an excellent discussion on the pathways of water loss from cut forages. Drying is faster in alfalfa than in most other legumes, but Thomas et al. (1983) reported that smooth bromegrass (*Bromus inermis* Leysser) dried faster than alfalfa. Legumes often contain more water than grasses because of a greater cell soluble fraction (Dougherty, 1987). Protoplasm can contain up to 95% water, while vacules may contain up to 98% water (Slayter, 1967). Owen and Wilman (1983) ranked the drying rates of various grasses as follows: tall fescue (*Festuca arundinacea* Schreber) > annual ryegrass = meadow fescue (*F. pratensis* Hudson) > timothy *Phleum pratense* L.) = orchardgrass (*Dactylis glomerata* L.) > perennial ryegrass. This ranking, however, would be greatly affected by the stage of development.

Accurately predicting forage physiological response to drying can be rather difficult since plant factors such as, species, maturity, temperature, plant organ, location within plant organ, and moisture level, and environmental factors such as, temperature, relative humidity, rainfall, and dew, can interact to cause postharvest changes. In some instances physiological change and potential nutrient losses can be significant, while in other situations they may be negligible.

PHYSIOLOGICAL CHANGES WITH DRYING

Immediately after cutting the plant remains alive with the stomata open. Nash (1959) and Clark et al. (1977) cite research that indicates a potential for additional photosynthesis by cut plants. Little or no additional dry matter is added by post-cutting photosynthesis because the canopy is no longer oriented effectively to intercept light and only the surface of the swathed forage is illuminated. Since stomatal activity is affected by light, leaf blades that are heavily shaded in the swath or windrow close their stomata rather quickly. Leaves that are exposed

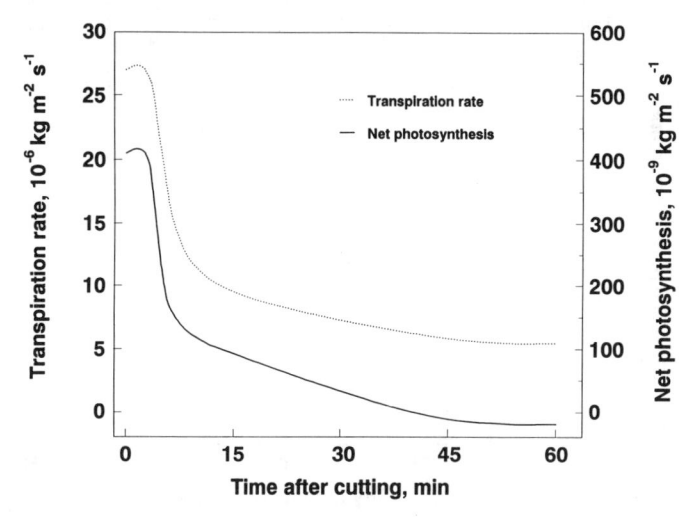

Fig. 1–2. Changes in net photosynthesis and transpiration rate after cutting annual ryegrass leaves (Clark et al., 1977).

to light lose moisture rapidly, which causes stomatal closure. Several authors indicate that stomata close within 1 to 2 h after cutting (Harris & Tullberg, 1980; Clark et al., 1977). In drying chamber studies, Clark et al.(1977) reported optical closure (appears closed to the eye) in as little as 15 min after cutting and a complete physiological closure (little to no gas exchange) after 30 to 40 min (Fig. 1–2). Johns (1972) reported that at 70% relative water content, stomatal closure was nearly complete for tall fescue, phalaris (*Phalaris aquatica* L.), and white clover (*Trifolium repens* L.). After cutting, Honig (1979) measured respiration activity in drying forage under light and dark conditions (Fig. 1–3). At 20% DM, photosynthetic CO_2 useage (under lighted conditions) reduced the respiratory CO_2 production by 50%. The reduction of respiratory loss by apparent photosynthesis ceased when the forage reached 30% DM. Apparently, photosynthesis was occurring in the early stages of drying. It could not add any net weight, however, only reduce the weight loss from respiration. Once transpiration ceases, the temperature of the mesophyll rises, increasing the metabolic activity of enzymes until they are denatured (Sullivan, 1969).

Care must be taken in applying literature from drought induced stress to changes that occur in cut plants. Intact plants that are drought stressed may osmotically adjust and continue to function physiologically. Osmotic adjustment may occur for a short time in cut plants, but the process may not have any significant consequences related to the forage. Abcisic acid (ABA) increases rapidly in wilted leaves and one of its most common effects is stomatal closure; however, stomatal closure generally precedes any rise in ABA concentration and occurs because of lack of turgor. Drought induced stress also may induce ethylene evolution too (Levitt, 1980).

Cell organelles vary in their response to dehydration. Chloroplasts and mitochondria have been reported to be severely damaged with drought stress, while peroxisomes were unaffected. In his book, Levitt (1980) cited work from Kurkova

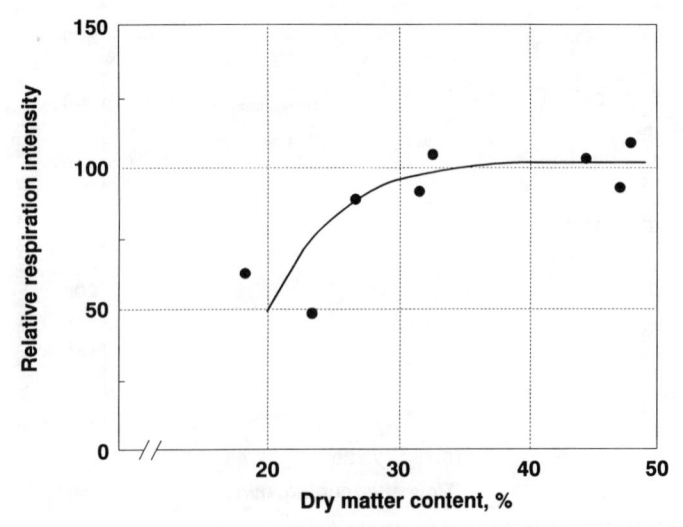

Fig. 1–3. The effect of photosynthetic compensation on relative respiration intensity as measured by respiratory CO_2 production. Respiration in the dark = 100 (Honig, 1979).

showing that thylakoid membranes swelled soon after leaves were detached from the plant and the lamellar system was disrupted. The stroma proteins crystallized 2 h after leaf detachment. When the tonoplast and plasmodesmata become nonfunctional and cytoplasmic membranes are disrupted, cell components are severely damaged and cells cannot recover with rehydration (Levitt, 1980). Shrinking of cell contents during drying and swelling during rehydration irreversibly damages cell membranes and plasmodesmata (Fitter & Hay, 1981). As the cells dry and the vacuoles shrink there is an inward pull on the protoplasm and an outward pull by the cell wall. As the protoplasm is torn, the membranes become physically damaged and further water loss occurs through physical processes (Sullivan, 1973).

Respiration

The post-harvest metabolic process with the most practical significance is respiration. Plant tissue continues to respire until the cells are no longer alive. The greatest change that occurs in drying is the respiration loss of carbohydrates and organic acids. This loss of readily digestible material makes even small respiratory losses important. Numerous researchers have looked at post-harvest respiratory losses in forage crops. Some studies show considerable loss, while others show relatively little. Honig (1979) compared mechanical losses with respiratory losses with different drying times and handling regimes (Fig. 1–4). Respiratory losses represented about one-quarter to one-third of the total losses, which represented 3 to 5% of the dry matter when packaged at 80% dry matter. Wilkinson (1981) summarized losses in ryegrass and white clover forage when harvested as nonwilted and wilted silage and as hay dried in the field (Table 1–1). Plant respiratory losses accounted for only 2 to 3% of the dry matter in wilted silage, but 8

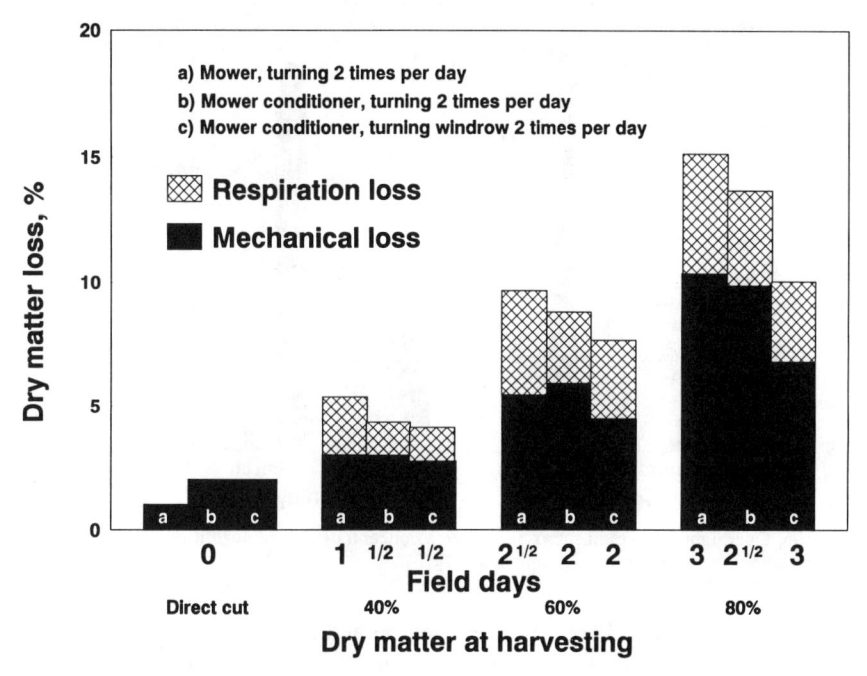

Fig. 1–4. Dry matter losses of grass during field drying (Honig, 1979).

to 9% in field dried hay. These respiratory losses represented ≈14 and 33% of the total dry matter losses for wilted silage and field dried hay, respectively. Respiratory losses ranging from of 2 to 8% are often quoted in the literature (Klinner & Shepperson, 1975; Melvin & Simpson, 1963). Under poor drying conditions, however, respiratory losses may be as high as 16% (Klinner & Shepperson, 1975). Knapp et al. (1973) observed an overnight decline in starch in both cut and uncut alfalfa in May (Fig. 1–5). With uncut alfalfa, translocation to the bases and root system could occur, but with cut forage the loss would be due to starch degrada-

Table 1-1. Typical dry matter (DM) losses of silage and hay under good management (Wilkinson, 1981).

Loss, %DM	Silage No Wilt	Silage Wilt	Field-dried hay
Field			
Respiration	--	2	8
Mechanical	1	4	14
Storage			
Respiration	--	1	1
Fermentation	5	5	2
Effluent	6	--	--
Surface waste	4	6	2
During removal	3	3	1
Total	19	21	28

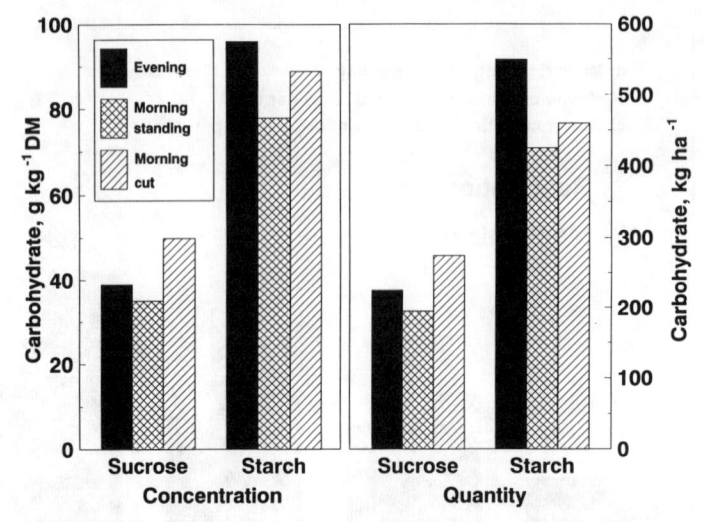

Fig. 1–5. Overnight changes in sucrose and starch concentrations and quantities per hectare in cut and standing alfalfa in May (Knapp et al., 1973).

tion. Sucrose increased both in concentration and quantity overnight reflecting the starch breakdown. These authors also indicated that the dry matter loss in cut alfalfa was greatly influenced by night temperature (Fig. 1–6), with dry matter losses doubling from ≈3 to 18°C. Under unfavorable drying conditions the ef-

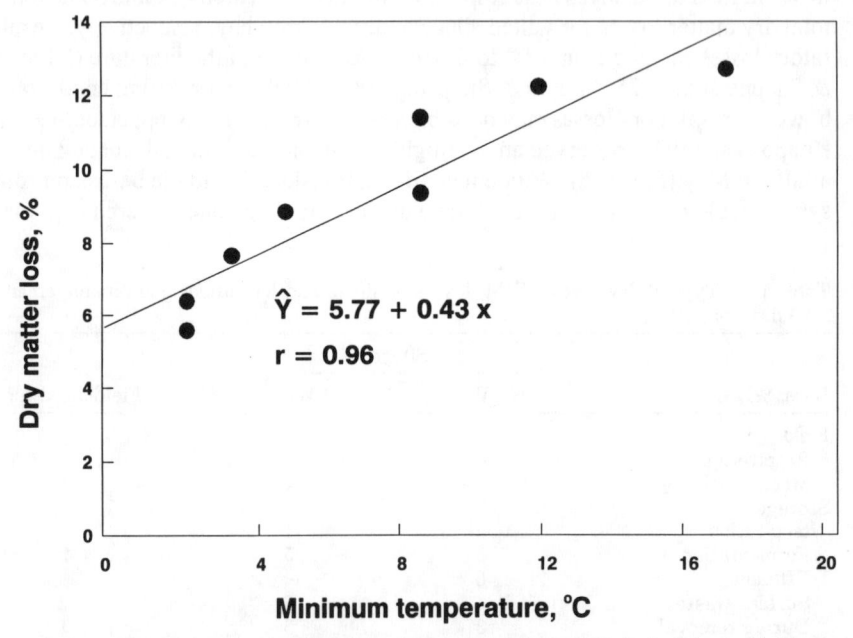

Fig. 1–6. Relationship between overnight dry matter loss in cut alfalfa and minimum night temperature (°C), July 1970 and May 1971 (Knapp et al., 1973).

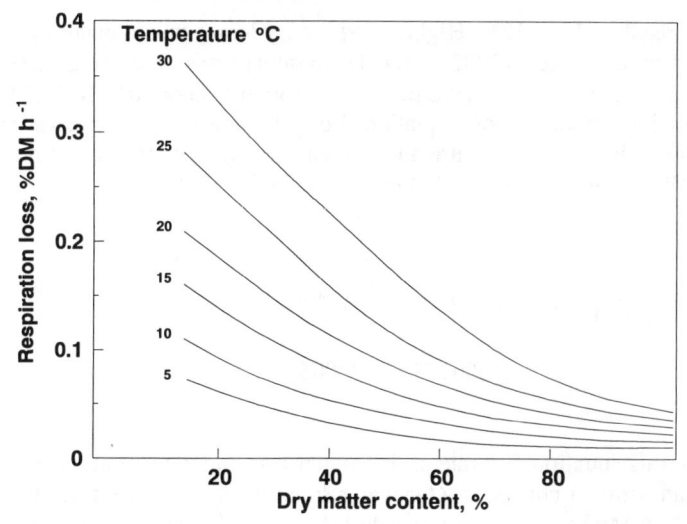

Fig. 1–7. Rate of respiration loss at six temperatures as affected by dry matter content (Honig, 1979).

fects of plant and microbial respiration are difficult to separate. During rainy weather some of the losses attributed to leaching may come from an extension of plant respiration and the beginning of microbial respiration (Honig, 1979).

Soluble carbohydrate loss from fresh to dried forage may be underestimated in some cases depending on how samples are dried. Burns et al. (1964) compared oven-dried alfalfa samples to freeze-dried samples at four dates. They found an average of five percentage points difference in soluble carbohydrates between freeze-dried samples and oven-dried samples at 77°C. In many studies soluble carbohydrate levels in the fresh sample may be underestimated, thus underestimating the loss of soluble carbohydrate. Wolf and Carson (1973) reported that respiration in alfalfa herbage was inactivated by tissue temperature above 55°C for 15 min and desiccation to ≈60% DM. A microwave treatment of 3 s at 1.0 kW reduced respiration by 71% in alfalfa leaves and by >95% with a 12-s treatment. In stems, respiration was completely eliminated with a 12-s treatment at 0.25 or 1.0 kW of energy. With a 12-s treatment 16 to 42% of the water was lost (Seif et al., 1983).

Honig (1979) related respiratory activity to temperature and dry matter content (Fig. 1–7). Respiratory rate decreased in a quadratic fashion at all temperatures as moisture was lost and the rate was directly related to temperature. The decline in respiration rate may be partially explained by the O_2 supply. When cells are in a flaccid state stomata are closed. There is less intercellular space and there is a greater area of contact among cells. This decreases the normal area for gas flow causing slower diffusion thus increasing the mesophyll resistance for O_2 (Levitt, 1980). Respiration ceases in a plant when the dry matter reaches 35 to 40% (dry basis) or at ≈25 to 30% if calculated on a wet basis (Greenhill, 1959; Klinner & Shepperson, 1975). Wood (1972), using data from Pizarro and James (1972), calculated that respiration would not completely cease until tissue water content reached 16 to 18%. Highest respiratory losses would occur under warm, humid conditions. Rees (1982) derived respiratory losses averaging 5.2% at 15°C, 7.2% at 20°C, and 9.4% at 25°C using data from the literature. McGechan (1989) developed a respiratory loss equation (Eq. [1]) for a forage conservation model relating respiration to moisture and temperature that fits the values reported by others (Rees, 1982; Wood & Parker, 1971; Honig, 1979).

$$L_r = a \left\{ \frac{Q}{K_m + Q} \right\} \{1 - 0.01b(D_e - D_c)\} \, e^{0.069T}$$

$$(0.000128m^2 - 0.00588m + 0.106) \qquad [1]$$

In this equation L_r = rate of dry matter loss due to respiration (% DM h^{-1}); a, b = adjustment constants, Q = water soluble carbohydrate concentration (% DM), K_m = Michaelis-Menten constant, D_e = D value at heading date (assumed 70%), D_c = D value at cutting, and m = moisture content on a wet basis.

As a general rule, respiration rates are greatest in young or meristematic tissues (Pizarro & James, 1972). They estimated that respiratory losses were between 12.8 g kg^{-1} when the inflorescences were emerging and 2.6 g kg^{-1} 30 d after anthesis. Greenhill (1959) found that the maturity of plants did not clearly affect respiration rates at various moisture contents; however, the higher moisture content and longer drying times of young grass increases the respiratory losses compared with more mature tissue. In laboratory experiments, crushing the stems of alfalfa increased the respiration rate compared with uncrushed stems, but not in proportion to the extent of crushing (Simpson, 1961). In some cases crushed stems had a respiration rate 15% higher then uncrushed stems. Crushing would not necessarily translate to greater respiratory loss since crushed stems dry more quickly than uncrushed ones so respiration would cease more quickly.

Carbohydrates comprise the largest amount of substrate for respiration and they are the primary substrate metabolized during drying (Parkes & Greig, 1974). Melvin and Simpson (1963) reported that fructans decreased sharply with drying and that fructose residues were rapidly respired. These substances accounted for most of the respiratory loss in ryegrass. Sucrose decreased early in the wilting process, but increased later in the drying cycle possibly as a result of synthesis from glucose and fructose in the plant. There was no trend in glucose content during the drying process indicating that breakdown of more complex carbohydrates and interconversions kept the glucose level constant. Melvin and Simpson, (1963) found that hexose sugars accounted for 32, 78, and 72% of the total CO_2 produced by air drying ryegrass plants at the booting, head emergence, and full flower stages, respectively. Organic acid changes amounting up to 10 g kg^{-1} of the DM may have occurred too. Rapid drying minimizes the breakdown and loss of respiratory substrates, but when environmental conditions prolong the drying period and the plant exhausts available carbohydrates many different kinds of compounds are respired including proteins.

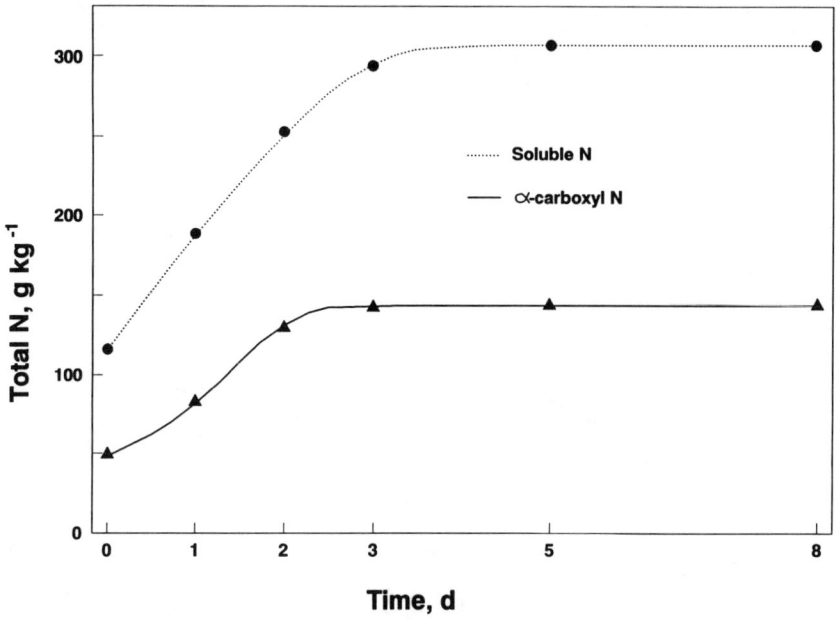

Fig. 1-8. Soluble N and α-carboxyl N in perennial ryegrass during 8 d of wilting (Kemble & Macpherson, 1954).

Proteolysis and Nitrogen Compounds

The literature on protein and N changes in plant material during the drying process is quite variable, probably due to the variable conditions that affect drying. Melvin and Simpson (1963) indicated that some protein breakdown was observed when drying ryegrass, but the major source of losses were soluble carbohydrates. Proteolysis during wilting of ryegrass produces peptides, amino acids, amides, and volatile base (ammonium; Kemble & Macpherson, 1954; Fig. 1-8 and 1-9). They reported that fresh ryegrass had 5 g kg^{-1} of the total N as peptides in the soluble N fraction. This increased to 25 and 33 g kg^{-1} after 1 and 8 d of drying, respectively. The concentration of soluble N related to total N increased sharply for 3 d after cutting. Likewise, the α-carboxyl-N increased for 3 d indicating breakdown of proteins to peptides (Fig. 1-8). Total amide N increased little during the first day after cutting in plants with <20% DM, but total amide N and the ammonium level rose modestly from Day 1 to Day 3 when the DM rose from <20% to ≈60% (Fig. 1-9). In this study they looked at changes in amino acid composition during 8 d of wilting. Glycine, alanine, tyrosine, and leucine were all at lower values during the 8 d of drying than would be expected with a uniform breakdown of protein. Concentrations of serine threonine, valine, methionine, and phenylalanine in wilted ryegrass were only slightly less that would be expected with a uniform breakdown of protein. Kemble and Macpherson (1954) were the first authors to find greatly elevated levels of proline in water stressed leaves (Fig. 1-10). They found up to five times more proline than would be nor-

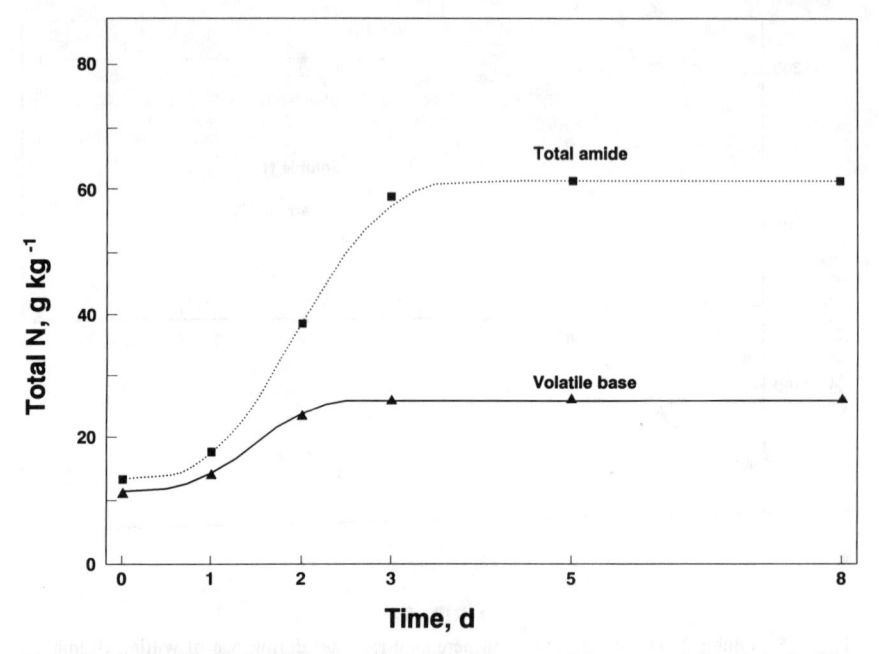

Fig. 1–9. Total amide and volatile base (ammonium) in perennial ryegrass during 8 d of wilting (Kemble & Macpherson, 1954).

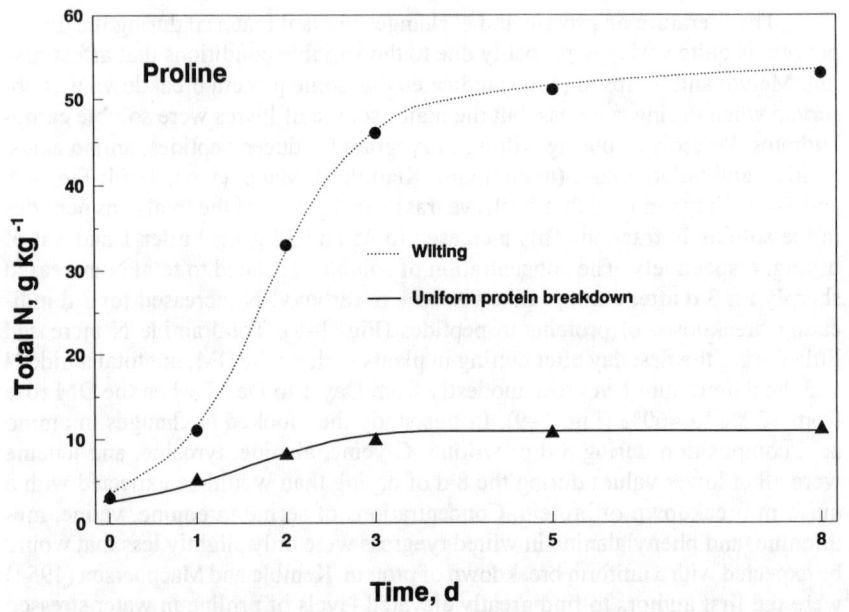

Fig. 1–10. Comparison of proline formed in perennial ryegrass during 8 d of wilting with the amount calculated assuming normal breakdown of protein (Kemble & Macpherson, 1954).

Table 1-2. The effect of wilting on the major N components of ryegrass-clover (Carpintero et al., 1979).

	Dry matter content	Protein N	Ammonia N
	g kg^1	— g kg^{-1} total N —	
Fresh grass	173 ± 1.3	925 ± 8.5	1.2 ± 0.07
Rapidly wilted grass, 6 h	349 ± 3.0	878 ± 6.8	1.1 ± 0.03
Rapidly wilted grass, 48 h	462 ± 13.5	832 ± 9.1	2.1 ± 0.06
Moist wilted grass, 48 h	199 ± 4.6	752 ± 9.9	2.6 ± 0.12
Moist wilted grass, 144 h	375 ± 35.4	689 ± 34.4	26.1 ± 2.45

mally present with uniform protein breakdown. During the first 3 d, proline content increased markedly and then at a much slower rate afterward. In identical grass samples that were not allowed to lose moisture during the starvation process proline was not synthesized. Proline not only comes from protein and peptide breakdown, but from de novo synthesis as well (Levitt, 1980). Proline and amide production could be one way that plants conserve N during amino acid metabolism and deamination. Proline accumulation and metabolism in relation to water stress is discussed by Stewart and Hanson (1980).

Minimum protein breakdown and ammonium production occurred with rapid wilting (Table 1–2; Carpintero et al., 1979). Ammonium did not appear in large quantities until after plants are kept alive for an extended period of time. Spoelstra and Hindle (1989) evaluated field wilted ryegrass prior to harvesting as silage. The average increase in NH$_3$–N was 5.8 g kg^{-1} N d^{-1}, which occurred under rain. They found extensive protein degradation during wilting and found no relationship between lengths of wilting and total N or ash content. Ammonium formed in the field comes from deamination of amino acids and amides and it is difficult to separate that caused by plant enzymes from that caused by microorganisms. Papadopoulos and McKersie (1983) examined protein hydrolysis during wilting and ensiling of alfalfa, red clover (*T. pratense* L.), birdsfoot trefoil (*Lotus corniculatus* L.), smooth bromegrass, and timothy. In the 24-h wilting period for both the first and second cuts, protein was hydrolyzed to soluble nonprotein N (SNPN) to the greatest extent in alfalfa and to the least extent in red clover. At first cut, herbage of various species contained 40 to 90 g kg^{-1} total N as SNPN. After wilting, this level increased to 110 to 250 g kg^{-1}. Alfalfa consistently had the highest amount of proteolysis and red clover the least. Papadopoulos and McKersie (1983) indicated that proteinase activity was highest in direct-cut herbage and decreased as forage dried in the legume and grass species they studied.

Bloat rarely occurs with dry legumes so the change in protein configuration and its breakdown apparently reduced the foam forming properties of soluble proteins (Sullivan, 1969). Nitrates appear to be affected very little in the drying process (Sullivan, 1973). Nitrate reductase activity decreases rapidly with drying so nitrates are not metabolized to any extent (Levitt, 1980). Interconversions of N compounds occur with drying, but most studies show that there is little metabolic loss of N.

Enzyme Systems

Enzymatic activity continues after forage is cut. Intrinsic hydrolysis and respiration continues until cells lose their integrity (Sullivan, 1969). McDonald (1973) indicated that plant enzyme activity continues when green or wilted material is put into a silo as long as aerobic conditions exist. This would contribute to heat production and plant enzymatic activity then would cease in a matter of hours. Upon ensiling, aerobic microbial activity is difficult to separate from plant metabolism. Drying of cells causes direct damage to enzyme systems. Air drying forages probably inactivates nearly all enzymes (Todd, 1972). As water associated with proteins is removed conformational changes take place in enzymes. Inactivation may be caused by the formation of intra- or intercellular disulfide bonds. Drying may lead to the activation of degradative enzymes. Maintenance of membrane integrity is necessary in order for the cell to continue living. Once membrane integrity is lost, irreversible interactions can occur among contents of various compartments (Todd, 1972). Leaves usually die after 40 to 90% of the total water is lost (Todd & Yoo, 1964). Perennial ryegrass leaves have been shown to die at -1.5 MPa (Sheehy et al., 1975) and wheat (*Triticum aestivum* L.) leaves died at -3.5 to -4.0 MPa (Barlow et al., 1977).

Todd and Yoo (1964) followed enzymatic changes in detached wheat leaves that were held in the dark either dried over desiccants or held over water to maintain leaf water content. The leaves lost ≈50% of their water after 16 h over the desiccants and were air dry after 48 h, which would simulate excellent drying conditions. Saccharase (invertase) showed the greatest sensitivity of any enzymes studied. Fifty percent of the saccharase activity was lost in 8 h or less if detached leaves were held over water (turgid samples), while 16 h of desiccation was required before 50% loss of saccharase activity occurred. Saccharase activity was low in both turgid and dried samples 25 h after leaf removal. This decrease in saccharase activity accounts for the two-fold increase of sucrose observed in wheat leaves subjected to drought. In general, phosphatase activity did not decrease as rapidly, but phosphatase activity in desiccated leaves dropped more rapidly and to a lower level than in turgid leaves. Desiccated leaves lost 50% of their phosphatase activity in 24 h, but turgid samples maintained phosphatase activity above 60% for 48 h. Peroxidase activity acted much differently. Peroxidase activity dropped slowly in the desiccated leaves and 50% of the activity remained when the leaves were air dry (3% moisture). In turgid samples there was a small initial drop in peroxidase activity and then a large increase. After 48 h the peroxidase activity was 150% that of the initial value. Peptidase activity did not drop when leaves were held over water and only dropped ≈20% in desiccated samples in 48 h. Dehydrogenase activity only was slightly affected in the turgid samples, dropping ≈20% in 48 h. In the desiccated samples it increased to ≈120% of the initial value during the first 20 h and then activity dropped off sharply to <60% 25 h after commencement of drying, followed by a gradual decline to 50% of the original activity by 48 h. The protein content declined rapidly with desiccation to ≈50% of its original value after 40 h. In turgid samples, protein dropped to 34% after 48 h.

Other Compounds

Vitamins

Carotene (Vitamin A precursor) is the most easily destroyed nutrient in a forage crop. Losses range from 90 to 95% in field cured hay, 80 to 90% in barn-dried hay, and 40 to 60% in silage (Carter, 1960). Sullivan (1969) indicated that there is a major loss of carotene in drying due to lipoxidase destruction. The loss may be especially high on hot days. Slow drying in hot (37°C) humid conditions maximizes carotene destruction (Sullivan, 1973). Photo-oxidation of carotene may occur with direct sunlight or ultraviolet light or light may just increase tissue temperature and lipoxidase activity (Sullivan, 1969). Rapid drying, either naturally or artificially, quickly inactivates lipoxidase and reduces carotene loss (Sullivan, 1973).

Vitamin E (tocopherols) are highest in young plants and lowest in mature plants. Vitamin E content also is reduced by drying. Tocopherol levels in blood of animals on pasture were higher than those receiving stored forages (Sullivan, 1973). Vitamin D is present in very small quantities in green forages, but Vitamin D is produced when various sterols are irradiated in partially dried or dried plant cells (Sullivan, 1973).

Antimetabolites

Prussic Acid. Dhurrin, the cyanogenic glucoside containing HCN (prussic acid) is located in the epidermis, while the enzymes responsible for dhurrin catabolism and HCN release are almost exclusively located in the mesophyll (Kojima et al., 1979). Rapid release of cyanide occurs when plant tissue is crushed (with livestock consumption) and dhurrin is mixed with endogenous enzymes or rumen microflora (Tapper & Reay, 1973). Generally, it is thought that plants that contain cyanogenic glucosides lose their toxicity when dried. It has been suggested that this is a result of denaturation of the enzyme systems causing the release of HCN, volatilization of the free cyanide, or both (Sullivan, 1973); however, Haskins et al. (1988) oven-dried sorghum [*Sorghum bicolor* (L.) Moench] leaf tissue at 75°C for 2 h and maintained the concentration of dhurrin. They concluded that leaf tissue could be oven dried at 65 to 85°C without loss of HCN_p (HCN potential). When the cells lose their integrity the enzymes and substrate can mix releasing HCN (Kojima et al., 1979). With rapid oven drying, however, by the time the cell membranes lose their integrity there may be insufficient water remaining to promote dhurrin hydrolysis, or enzymes that release HCN may be denatured. Therefore, prussic acid would be more likely to dissipate in drying following a freeze than from rapid drying of relatively undisturbed plant leaf tissue. Black cherry (*Prunus serotina* Ehrh.) had lower levels of HCN_p in leaves air dried for 24 h than in fresh leaves. Leaves that were air dried for 48 h had very little HCN_p even when immature (Smeathers et al., 1973).

Alkaloids. Alkaloids only are affected slightly by the drying process. Perloline content in tall fescue was changed very little by oven drying compared with freeze drying, but perloline content was reduced by field drying (Culvenor,

1973). Candrian et al. (1984) reported that the pyrrolizidine alkaloids (PA) in a groundsel (*Senecio alpinus* L.) remained at a constant level in hay containing the plant; however, when the groundsel was mixed with silage much of the PA were destroyed in the ensiling process. As much PA was retained by air drying for 4 d and then oven drying as with immediate oven drying at 60°C. Steers (*Bos taurus*) fed endophyte (*Acremonium coenophialum* Morgan-Jones & Gams) infected tall fescue hay exhibited elevated temperatures, had lower forage intake, and had lower daily gains compared with animals fed endophyte-free tall fescue hay (Schmidt et al., 1982). There was a high concentration of loline alkaloids in endophyte-infected tall fescue hay (Jackson et al., 1984). They reported tall fescue toxicosis symptoms (elevated body temperature and respiration rate) with calves fed endophyte infected fescue hay. Their endophyte-infected tall fescue hay retained its toxicity after dehydration and 3 yr of storage. Drying high alkaloid material apparently does not eliminate the toxicity.

Tannin. Terrill et al. (1989) compared tannin concentration in fresh frozen and field dried forage in both high tannin and low tannin types of serecia lespedeza [*Lespedeza cuneata* (Dum.-Cours.) G. Don] (Table 1–3). In high tannin lines, the tannin level of fresh frozen samples of serecia lespedeza contained 181 g kg^{-1} DM tannin. This was reduced to 31 g kg^{-1} with field drying. In low tannin lines, serecia lespedeza herbage contained 87 g kg^{-1} DM tannin if freeze dried and 44 g kg^{-1} if field dried. Drying would be a practical way to increase the consumption of high tannin serecia lespedeza forage. Terrill et al. (1989) have observed that cattle (*Bos taurus*) readily consume field dried high tannin serecia lespedeza hay, while their intake of the fresh material in pasture is low.

REWETTING

If rewetting occurs before irreversible metabolic or structural damage to drying forages takes place, metabolic activity may resume and respiratory losses will continue. After membranes are no longer functional electrolyte leakage may occur with added water (Levitt, 1980). After the plant material is no longer living, rewetting can increase microbial activity and substrate utilization. Collins (1983) reported that losses subsequent to rewetting may result from leaf loss, leaching, and respiration. Rewetting reduced total nonstructural carbohydrates in both alfalfa and red clover. In vitro dry matter digestibility was consistently reduced and lignin content increased with rewetting. Neutral detergent fiber increased in red clover, but not in alfalfa. Rewetting did not change forage N concentration even though more than a 50% DM loss was reported in some situations.

Rain did not affect mineral content except for Ca in legume and legume–grass hays (Collins, 1985b). Sixty-two millimeters of precipitation reduced K concentration in alfalfa hay indicating that considerable leaching may have occurred. The P, Ca, and Mg concentrations were increased after rewetting (Collins, 1985a), which suggests that DM losses were greater than mineral nutrient losses.

SUMMARY

A considerable amount of nutrient loss may occur in the forage harvesting process. Much of it is due to mechanical leaf shatter. Plant metabolism after cutting may cause small losses under rapid drying conditions, but with prolonged drying the plant enters a starvation process and the metabolic losses may be significant. Respiration of carbohydrates and organic acids result in the greatest metabolic loss. Since these compounds are all highly digestible, even a small loss may have nutritional significance. Proteins may be broken down and respired, especially when cut plants remain alive during an extended drying period. Although there may be considerable protein breakdown and interconversions of N compounds, N is conserved and little appears to be lost due to metabolic processes under most drying conditions. Changes in plant composition attributed to leaching of soluble components may occur after cell membranes lose their integrity and the forage is wetted by precipitation. Leaching losses of organic compounds are difficult to separate from microbial induced losses. Antimetabolites such as HCN_p and tannin are reduced by drying, but alkaloid concentrations are affected very little.

REFERENCES

Albrecht, K.A., W.F. Wedin, and D.R. Buxton. 1987. Cell-wall composition and digestibility of alfalfa stems and leaves. Crop Sci. 27:735–741.

Barlow, E.W.R., R. Munns, N.S. Scott, and A.H. Reisner. 1977. Water potential, growth, and polyribosome content of the stressed wheat apex. J. Exp. Bot. 28:909–916.

Burns, J.C., C.H. Noller, and C.L. Rhykerd. 1964. Influence of method of drying on the soluble carbohydrate content of alfalfa. Agron. J. 56:364–365.

Buxton, D.R. 1990. Cell wall components in divergent germplasms of four perennial forage grass species. Crop Sci. 30:402–408.

Candrian, V., J. Lüthy, P. Schmid, C. Schlatter, and E. Gallasz. 1984. Stability of pyrrolizidine alkaloids in hay and silage. J. Agric. Food Chem. 32:935–937.

Carpintero, M.C., A.R. Henderson, and P. McDonald. 1979. The effect of some pre-treatments on proteolysis during ensiling of herbage. Grass Forage Sci. 34:311–315.

Carter, W.R.B. 1960. A review of nutrient losses and efficiency of conserving forage as silage, barn dried hay and field cured hay. J. Brit. Grassl. Soc. 15:220–230.

Clark, B.J., J.L. Prioul, and H. Couderc. 1977. The physiological response to cutting in Italian ryegrass. J. Brit. Grassl. Soc. 32:1–15.

Collins, M. 1983. Wetting and maturity effects on the yield and quality of of legume hay. Agron. J. 75:523–529.

Collins, M. 1985a. Wetting and maturity effects on mineral concentrations in legume hay. Agron. J. 77:779–782.

Collins, M. 1985b. Wetting effects on the yield and quality of legume and legume-grass hays. Agron. J. 77:936–941.

Culvenor, C. C. 1973. Alkaloids. p. 375–446. *In* G.W. Butler and R.W. Bailey (ed.) Chemistry and biochemistry of herbage. Vol. 1. Academic Press, New York.

Dougherty, C.T. 1987. Post-harvest physiology and preservation of forages. p. 12–20. *In* Proc. Am. Forage and Grassl. Council, Springfield, IL. 2–5 Mar. 1987. Am. Forage and Grassl. Council, Lexington, KY.

Fitter, A.H., and R.K.M. Hay. 1981. Environmental physiology of plants. Academic Press, New York.

Greenhill, W.L. 1959. The respiration drift of harvested pasture plants during drying. J. Sci. Food Agric. 10:495–501.

Harris, C.E., and M.S. Dhanoa. 1984. The drying of component parts of inflorescence-bearing tillers of Italian ryegrass. Grass Forage Sci. 39:271–275.

Harris, C.E., and J.N. Tullberg. 1980. Pathways of water loss from legumes and grasses cut for conservation. Grass Forage Sci. 35:1–11.

Haskins, F.A., H.J. Gorz, and R.M. Hill. 1988. Colorimetric determination of cyanide in enzyme-hydrolyzed extracts of dried sorghum leaves. J. Agric. Food Chem. 36:775–778.

Holt, D.A., and A.R. Hilst. 1969. Daily variation in carbohydrate content of selected forage crops. Agron. J. 61:239–242.

Honig, H. 1979. Mechanical and respiration losses during pre-wilting of grass. p. 201–204. In C. Thomas (ed.) Forage conservation in the 80's. Occasional symp. no. 11. Brit. Grassl. Soc. Janssen Services, London.

Jackson, J.A., R.W. Hemken, J.A. Boling, R.J. Harmon, R.C. Buckner, and L.P. Bush. 1984. Loline alkaloids in tall fescue hay and seed and their relation to summer fescue toxicosis in cattle. J. Dairy Sci. 67:104–109.

Johns, G.G. 1972. Water use efficiency in dryland herbage production. Thesis summary. J. Aust. Inst. Agric. Sci. 38:135–136.

Jones, L. 1979. The effect of stage of growth on the rate of drying of cut grass at 20°C. Grass Forage Sci. 34:139–144.

Jones, L., and C.E. Harris. 1979. Plant and swath limits to drying. Forage conservation in the 80's. Occasional symp. no. 11. Brit. Grassl. Soc. Janssen Services, London.

Kemble, A.R., and H.T. Macpherson. 1954. Liberation of amino acids in perennial rye grass during wilting. Bioch. J. 58:46–49.

Klinner, W.E., and G. Shepperson. 1975. The state of haymaking technology--A review. J. Brit. Grassl. Soc. 30:259–266.

Knapp, W.R., D.A. Holt, V.L. Lechtenberg, and L.R. Vough. 1973. Diurnal variation in alfalfa (*Medicago sativa* L.) dry matter yield and overnight losses in harvested alfalfa forage. Agron. J. 65:413–417.

Kojima, M., J.E. Poulton, S.S. Thayer, and E.E. Conn. 1979. Tissue distributions of dhurrin and of enzymes involved in its metabolism in leaves of *Sorghum bicolor*. Plant Physiol. 63:1022–1028.

Levitt, J. 1980. Responses of plants to environmental stresses. Vol. II. Water, radiation, salt, and other stresses. Academic Press, New York.

Lyttleton, J.W. 1973. Proteins and nucleic acids. p. 63–103. In G.W. Butler and R.W. Bailey (ed.) Chemistry and biochemistry of herbage. Vol. 1. Academic Press, New York.

Macdonald, A.D., and E.A. Clark. 1987. Water and quality loss during field drying of hay. Adv. Agron. 41:407–437.

McDonald, P. 1973. The ensilage process. p. 33–60. In G.W. Butler and R.W. Bailey (ed.) Chemistry and biochemistry of herbage. Vol. 3. Academic Press, New York.

McGechan, M.B. 1989. A review of losses arising during conservation of grass forage. Part 1. Field losses. J. Agric. Eng. Res. 44:1–21.

Melvin, J.F., and B. Simpson. 1963. Chemical changes and respiratory drift during air drying of ryegrass. J. Sci. Food Agric. 14:228–234.

Minson, D.J., W.F. Raymond, and C.E. Harris. 1960. Studies in the digestibility of herbage. VIII. The digestibility of S37 cocksfoot, S23 ryegrass, and S24 ryegrass. J. Brit. Grassl. Soc. 15:174–180.

Murdoch, J.C. 1980. The conservation of grass. p. 174–215. In W. Holmes (ed.) Grass--Its production and utilization. Blackwell Scientific Publ., Oxford.

Nash, M.J. 1959. Partial wilting of grass crops for silage. J. Brit. Grassl. Soc. 14:65–73.

Nash, M.J. 1985. Crop conservation and storage. 2nd ed. Pergamon Press, Oxford.

Owen, I.G., and D. Wilman. 1983. Differences between grass species and varieties in rate of drying at 25°C. J. Agric. Sci. (Cambridge) 100:629–636.

Papadopoulos, Y.A., and B.D. McKersie. 1983. A comparison of protein degradation during wilting and ensiling of six forage species. Can. J. Plant Sci. 63:903–912.

Parkes, M.E., and D.J. Greig. 1974. The rate of respiration of wilted ryegrass. J. Agric. Eng. Res. 19:259–263.

Pizarro, E. A., and D.B. James. 1972. Estimates of respiratory rates and losses in cut swards of *Lolium perenne* (S321) under simulated hay making conditions. J. Brit. Grassl. Soc. 27:17–21.

Rees, D.V.H. 1982. A discussion of sources of dry matter loss during the process of hay making. J. Agric. Eng. Res. 27:469–479.

Schmidt, S.P., C.S. Hoveland, E.M. Clark, N.D. Davis, L.A. Smith, H.W. Grimes, J.L. Holliman. 1982. Association of an endophytic fungus with fescue toxicity in steers fed Kentucky 31 tall fescue seed or hay. J. Anim. Sci. 55:1259–1263.

Seif, S.A., D.A. Holt, V.L. Lechtenberg, and R.J. Vetter. 1983. Effects of microwave treatment on drying and respiration in cut alfalfa. p. 639–642. *In* J.A. Smith and V.W. Hays (ed.) Proc. 14th Int. Grassl. Cong., Lexington, KY. 15–21 June 1981. Westview Press, Boulder, CO.

Sheehy, J.E., R.M. Green, and M.J. Robson. 1975. The influence of water stress on the photosynthesis of a simulated sward of perennial ryegrass. Ann. Bot. (London) 39:387–401.

Simpson, B. 1961. Effect of crushing on the respiratory drift of pasture plants during drying. J. Sci. Food Agric. 12:706–712.

Slayter, R.O. 1967. Plant water relationships. Academic Press, New York.

Smeathers, D.M., E. Gray, and J.H. James. 1973. Hydrocyanic acid potential of black cherry leaves as influenced by aging and drying. Agron. J. 65:775–777.

Smith, D. 1971. Efficiency of water for extraction of total nonstructural carbohydrates from plant tissue. J. Sci. Food Agric. 22:445–447.

Smith, D. 1973. The non-structural carbohydrates. p. 105–155. *In* G.W. Butler and R.W. Bailey (ed.) Chemistry and biochemistry of herbage. Vol. 1. Academic Press, New York.

Spoelstra, S.F., and V.A. Hindle. 1989. Influence of wilting on chemical and microbial parameters of grass relative to ensiling. Neth. J. Agric. Sci. 37:355–364.

Stewart, C.R., and A.D. Hanson. 1980. Proline accumulation as a metabolic response to water stress. p. 173–189. *In* N.C. Turner and P.J. Kramer (ed.) Adaptation of plants to water and high temperature stress. John Wiley & Sons, New York.

Sullivan, J.T. 1969. Chemical composition of forages with reference to the needs of the grazing animal. USDA-ARS Bull. 34-107 USDA-ARS, Washington, DC.

Sullivan, J.T. 1973. Drying and storing herbage as hay. p. 1–31. *In* G.W. Butler and R.W. Bailey (ed.) Chemistry and biochemistry of herbage. Vol. 3. Academic Press, New York.

Tapper, B.A., and P.F. Reay. 1973. Cyanogenic glycosides and glucosinates. p. 447–476. *In* G.W. Butler and R.W. Bailey (ed.) Chemistry and biochemistry of herbage. Academic Press, New York.

Terrill, T.H., W.R. Windham, C.S. Hoveland, and H.E. Amos. 1989. Forage preservation method influences on tannin concentration, intake, and digestibility of serecia lespedeza by sheep. Agron. J. 81:435–439.

Thomas, J.W., T.R. Johnson, M.A. Weighart, C.M. Hanson, M.B. Tesar, and Z. Helsel. 1983. Hastening hay drying. p. 645–648. *In* J.A. Smith and V.W. Hays (ed.) Proc. 14th Int. Grassl. Cong., Lexington, KY. 15–21 June 1981. Westview Press, Boulder, CO.

Todd, G.W. 1972. Water deficits and enzymatic activity. p. 177–216. *In* T.T. Kozlowski (ed.) Water deficits and plant growth. Vol. III. Plant responses and control of water balance. Academic Press, New York.

Todd, G.W., and B.Y. Yoo. 1964. Enzymatic changes in detached wheat leaves as affected by water stress. Phyton (Buenos Aires) 21:61–68.

Wilkinson, J.M. 1981. Losses in the conservation and utilization of grass and forage crops. Ann. Appl. Biol. 98:365–375.

Wolf, D.D., and E.W. Carson. 1973. Respiration during drying of alfalfa herbage. Crop Sci. 13:660–662.

Wood, J.G.M. 1972. Letter to the editor. J. Brit. Grassl. Soc. 27:193–194.

Wood, J.G.M., and J. Parker. 1971. Respiration during the drying of hay. J. Agric. Eng. Res. 16:179–191.

2 Microbiology of Stored Forages

Craig A. Roberts
University of Missouri
Columbia, Missouri

Forage plants growing in the field are naturally inoculated with a wide range of fungi and bacteria. Those fungi that dominate microbial populations in field conditions have been termed *field fungi* (Christensen & Kaufman, 1965; Magan & Lacey, 1987); they include genera such as *Alternaria, Cladosporium,* and *Fusarium* fungi that proliferate on host tissues containing large concentrations of water (Magan & Lacey, 1987). In contrast, genera referred to as *storage fungi,* such as *Aspergillus* and *Fusarium* (Magan & Lacey, 1987; Roberts et al., 1991), require less water for growth and proliferate on dryer tissues (Fig. 2–1).

All of these species, whether classified as field or storage fungi, produce a wide range of toxic metabolites (Cole & Cox, 1981). In addition to their production of mycotoxins, the presence of spores causes respiratory problems. In humans, these problems lead to a syndrome called farmer's lung (Lacey & Lord, 1977). In livestock, respiratory problems are usually not as severe, with the exception of colic in horses (*Equus caballus*; Hintz & Lowe, 1977). In addition to the toxic effects, fungi greatly reduce the nutritional quality of feed and forage (Jones et al., 1955), as is discussed in greater detail in another chapter (Jaster, 1995, this publication).

The relative proportions of these fungi, bacteria, and other microorganisms change as forage is clipped, cured, harvested, and stored. Throughout production processes, microbial demographics are influenced by a number of chemical and physical factors, especially those of the host plant and environment. This chapter will focus on the effects of water, temperature, and pH on microbial populations and discuss several analytical methods for analyzing fungi in contaminated hay.

MICROBIOLOGY OF HAY

Of the factors influencing microbial populations in hay production and storage, moisture and temperature of the plant may be regarded as most important. Both factors are interrelated (Magan & Lacey, 1987); yet they can independently and synergistically regulate which microbial species populate a bale of hay.

Copyright © 1995 Crop Science Society of Agronomy and American Society of Agronomy, 677 S. Segoe Rd., Madison, WI 53711, USA. *Post-Harvest Physiology and Preservation of Forages.* CSSA Special Publication no. 22.

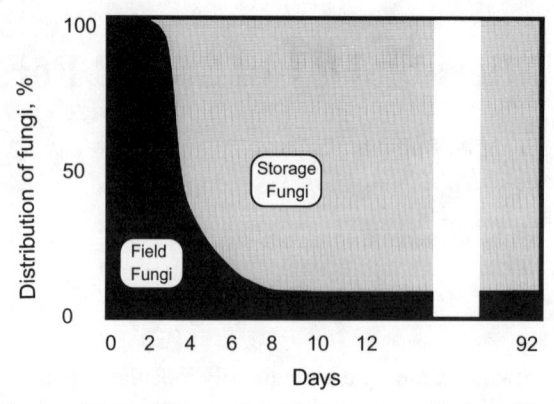

Fig. 2–1. Distribution of field and storage fungi during storage of untreated hay. From Kasperson et al., 1984. CFU, colony forming units.

Influence of Water on Microbial Populations in Hay

The species and proportion of microbes in hay are greatly influenced by moisture content. At higher moisture concentrations (>40%), plants continue to respire and microbial activity increases, resulting in higher temperatures. Because these effects are confounded, the moisture effect cannot be completely separated from the temperature effect; however, its general effect may be seen in those studies that have evaluated the effect of moisture on hay preservation. These effects can be seen in the series of experiments conducted at the Rothamsted Experiment Station from 1959 to 1962 (Gregory et al., 1963). In these experiments, hays were baled at three general moisture concentrations. Dry hays were baled at 16 to 17% moisture, normal wet hays at 25 to 28% moisture, and extremely wet hays at 39 to 42% moisture. Most of these hays were C-3 grasses, although some legumes were included as well.

In the first experiment, dry hays contained few spores of fungi or actinomycetes (10^6 g^{-1}). The normal wet hays, however, contained high concentrations of mesophilic fungi, primarily *Aspergillus glaucus*. Fungal spores in the normal wet hays reached a maximum concentration of 10^8 spore g^{-1} after 37 d of storage. They exhibited low concentrations of actinomycete spores, similar to those in the dry hays, rarely exceeding 10^6 g^{-1}. In the extremely wet hays, fungal spores reached 10^7 g^{-1} in 7 d and did not increase thereafter. Those fungal species increasing initially included *Absidia* sp., *Mucor pusillus, Aspergillus fumigatus*, and *Aspergillus nidulans*, thermophilic species that grow well between 40 and 60°C. The extremely wet hays developed very high levels of bacteria and actinomycetes, reaching 10^8 g^{-1} in 7 d.

Subsequent experiments in this study reported similar results, few actinomycetes, bacteria and fungal spores in dry hays, high numbers of fungal spores in normal wet hays, and high numbers of bacteria and actinomycete spores in extremely wet hays. When dry hay (17% moisture) was stored outside, it developed similar concentrations of fungal spores (near 2×10^7 g^{-1}) as hay baled at 65% moisture, though many fungi were not thermophilic. In addition, it only

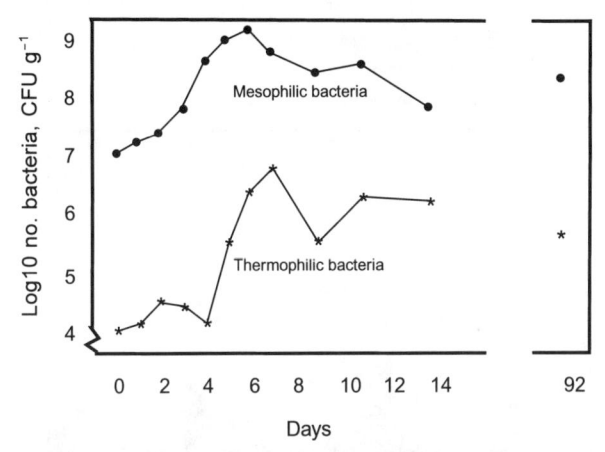

Fig. 2–2. Changes in mesophilic and thermophilic bacteria during storage. From Kasperson et al., 1984; CFU, colony forming units..

developed one-third the number of actinomycete and bacterial spores as the extremely wet hay.

When dry (15% moisture) and wet hay (30% moisture) was baled and stacked, there were few changes in the dry hay. This was not surprising, because there were relatively low populations of microorganisms in dry hay. In the wet hay, however, the stacked hay heated more than the baled hay. As could be expected, the uneven moisture concentrations throughout the stack resulted in uneven temperatures and produced uneven distributions of microbial species. The center of the stack contained high levels of *Bacillus licheniformis* and other spore-forming bacteria, but very few fungi; this portion of the bale was brown in color. Outside of this center were layers of gray, greenish-gray, and greenish-brown hay that contained large numbers of mold species.

Many other studies involving grass and legume hay spoilage report the presence of these same fungal and bacterial species (Lacey et al., 1978, 1981; Wittenberg & Moshtaghi-Nia, 1991; Woolford & Tetlow, 1984). In addition, other studies report similar changes in microbial populations as influenced by moisture content. Kaspersson et al. (1984) reported a shift from *Fusarium* and *Cladosporium* species to *A. flavus* and *A. glaucus* within the first 4 d of storage (Fig. 2–2). This shift, referred to earlier as a shift from field to storage fungi, occurred as water content decreased during 14 d of storage. They also reported a shift from gram-negative to gram-positive bacteria and attributed this to changes in moisture as well (Fig. 2–3).

Effect of Temperature on Microbial Populations in Hay

As was the case with moisture, the single effect of temperature on microbial populations in hay cannot be easily isolated. Its influence on microbial populations, however, can be seen in those studies detailing microbial changes at a constant moisture level. A good example is a study reported by Kaspersson et al.

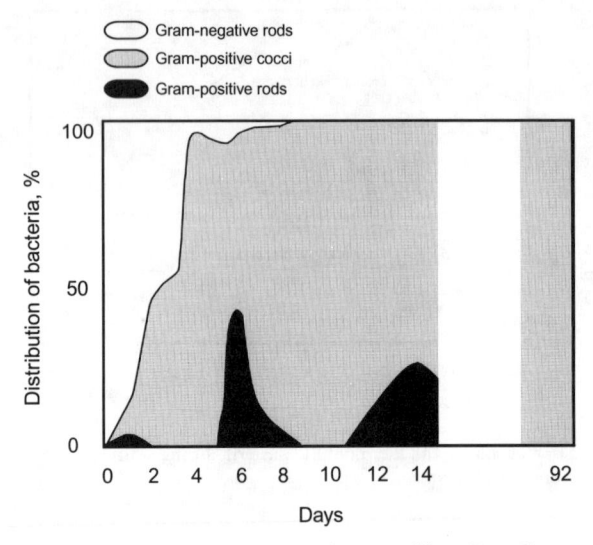

Fig. 2–3. Distribution of bacteria during storage of untreated hay. From Kasperson et al., 1984.

(1984). In this study, grass hay was baled at 31% moisture, and microbial changes were observed throughout the first 14 d of storage. Temperature rose quickly to 35°C, probably because of plant respiration. On Day 3, the temperature continued to rise possibly because of microbial activity. These increases in temperature were followed by changes in the ratio of mesophilic/thermophilic bacteria. Mesophilic bacteria initially increased until Day 6, when the population decreased to 10% of its maximum. Thermophilic bacteria, initially low in proportion to mesophilic bacteria, increased sharply from Day 4 to Day 7 and remained highly concentrated. The authors noted that the addition of urea did not affect the number of fungi and mesophilic bacteria; however, urea reduced temperature and numbers of thermophilic bacteria. The temperature of urea-treated hay did not exceed 40°C, while the temperature in control bales exceeded 50°C.

Corbaz et al. (1963) reported seven species of thermophilic and mesophilic actinomycetes in grass hay. They included *Micromonospora vulgaris, Thermopolyspora polyspora, T. glauca, Streptomyces thermoviolaceus, S. fradiae, S. griseoflavus,* and *S. olivaceus*. All seven species grew well at 40°C, and three were able to continue growing at 60°C. Woolford (1984) reported that mesophilic actinomycetes *S. griseus* and *S. albus* grew between 25 and 40°C, while the thermophilic species growing at 55°C included *T. vulgaris* and *Micropolyspora faeni*.

METHODS TO ASSESS MICROBIAL CONTAMINATION IN HAY AND FEED

Assessing microbial contamination in stored forage is important in many research studies and extension programs. Most procedures used to assess contamination in forage and feed analysis are designed to estimate molds. Each of these procedures has its own distinct advantages and disadvantages.

Visual Estimation

The most commonly used procedure for fungal or mold contamination is visual estimation (Goering & Gordon, 1973; Jeffers et al., 1982). It involves assessing the level of mycelia and spores, then relating estimates to an arbitrary scale of contamination. A typical scale may range from 1 to 5, where 1 = no visible spores or mycelia and 5 = considerable spores or mycelia (Roberts et al., 1987a). Basically, the visual estimation is a field technique. Although not quantitative, visual estimation can result in reliable assessment of contamination, especially when hay is extremely moldy or clean.

Visual estimation of fungal contaminants, however, has several disadvantages. First, it does not distinguish between slight differences in contamination, limiting its usefulness in quantitative research experiments. Second, a visual estimation does not always distinguish fungal spores from dust, as is commonly found in dry red clover (*Trifolium pratense* L.) hay; failure to do so could affect the relative rankings of hay lots when they are marketed. Third, visual estimation is by nature subjective, and thus it leads to controversy in product testing. Fourth, visual assessment usually requires a large, nonground sample, making sample handling inconvenient for analytical laboratories. Finally, the estimates recorded in this procedure are often affected by color, even though color is not always related to contamination.

Tabulation Methods

Other methods for assessing fungal contamination, such as plate (Cherney et al., 1987; Gregory et al., 1963), filament (Howard, 1911), and spore counts (Lacey & Dutkiewicz, 1976) involve tabulation of colonies or microscopic tabulation of mycelial fragments and spores. A typical spore count involves washing a sample in an aqueous solution containing a surfactant or detergent, then recording numbers of spores in a series of aliquots with the use of a hemocytometer. This method is quantitative, objective, and convenient and has several advantages. One obvious advantage over visual estimations is that the spore count allows a technician to distinguish spores from dust. It also allows for general taxonomic classification.

The spore count, however, is more tedious than visual estimation. It requires repeated analysis of several subsamples, each subsample requiring examination of multiple aliquots. And depending on how the sample is prepared and the aliquots are removed, data may only include large or small spores, not both. Another disadvantage of the spore count is that its basis for accuracy depends on a correlation between number of spores and level of mold; this correlation is not always high, probably because sporulation and mycelial growth fluctuate independently. Finally, the spore count's advantage of permitting taxonomic classification may not be important since there are multitudes of other procedures designed explicitly for that purpose (Agarwal & Sinclair, 1987).

Chemical Procedures

Perhaps the most accurate methods for assessing fungal contamination of stored forages and feeds are chemical procedures. Chemical methods involve quantification of fungal products such as chitin and ergosterol, constituents that are used as markers for total mycelial dry matter. Chitin is an excellent marker for mold contamination in hay and grain. It is stable, thereby serving as a marker for forages and feeds stored over several months and possibly years. In theory, chitin would not accurately represent level of contamination because of inconsistent levels among fungal species and physiological development. Chitin concentrations vary with stage of mycelial development (Bishop et al., 1982; Jarvis, 1977; Plassard et al., 1982; Swift, 1973) and among fungal species (Roberts et al., 1991), being absent from some fungi, such as oomycetes (Wessels & Sietsma, 1981). In addition, chitin from insect exoskeleton could interfere with that from fungal cell walls, resulting in elevated estimations. Yet despite potential inconsistency and interference, chitin has accurately estimated mycelial contamination in a wide range of forage and feed products (Roberts et al., 1987a and 1991; Wittenberg et al., 1989). In most cases when chitin fails to correlate with visual estimations, the error results from visually discriminating between bales of hay that are moderately contaminated (Roberts et al., 1987a,b); this problem stems from the qualitative procedure, not from chitin estimation. Chitin was superior to spore count when correlated to visual estimations in a wide range of grass and legume hays (Roberts et al., 1987a). When chitin does not correlate well with spore count or mold count, most error is attributed to tabulated data, and chitin data are regarded as accurate (Cousin et al., 1984; Golubchuk et al., 1960; Jarvis, 1977).

A clear disadvantage of using chitin to estimate fungal contamination is tedious laboratory procedure. A common chitin procedure involves extraction, separation, and quantification procedures developed by Ride and Drysdale (1972) and Tsuji et al. (1969). Chitin is decollated and partially hydrolyzed with alkaline or enzymatic reagents, separated and precipitated with celite and ethanol and refrigerated centrifugation, and measured as an aldehyde of N-acetyl-D-glucosamine with a colorimetric assay. Although this procedure is accurate and widely used, it is perhaps the most tedious procedure for estimating fungal contamination; it requires more time to analyze a sample for chitin than to analyze it for the combined forage quality components of acid and neutral detergent fiber, moisture, and crude protein. The separation of chitosan involves one centrifugation for celite-mediated precipitation, followed by two or three centrifugations to rinse the pellet with aqueous ethanol. Centrifuging only 50 samples can occupy one technician for 3 h, and deleting the centrifuging steps results in interfering compounds that greatly increase the intensity of the chromophore during the assay (Jones et al., 1985). The colorimetric assay itself is equally time-consuming, especially if assay duplicates are prepared. The assay includes 10 to 15 steps and is conducive to error from particles suspended in the cuvette. Percentage of recovery ranges from 30 to 100%, depending on whether rates represent glucosamine recovered from crab shell chitin added at hydrolysis or from spiked glucosamine (Bethlenfalvay et al., 1981; Roberts et al., 1987a; Ride & Drysdale, 1972). Although this procedure remains popular, many of its cumbersome steps can be eliminated with newer chromatographic and detection techniques for glucosamine (Cheng & Boat, 1978; Hubbard et al., 1979; Lin & Cousin, 1985; Mawhinney, 1986; Mawhinney et al., 1980).

The fungal sterol ergosterol also is used as a chemical marker for mold contamination (Seitz et al., 1979; Seitz & Pomeranz, 1983). Ergosterol is believed to be present in nearly all fungi (Weete, 1974), although it probably varies among species and stages of development. Ergosterol procedures are simple and precise, and they offer the most promise in grain analysis. Ergosterol is probably less useful as a marker for mold in hay. Although not yet reported, ergosterol concentrations probably increase initially in moldy hay, then decrease over long months of storage, especially in southern climates. This can be expected because ergosterol is easily oxidized. Future research may show that ergosterol accurately represents mold levels in hay, even when bales are stored outside through hot, dry summer months. But more than likely, studies will show that mycelia and ergosterol would accumulate simultaneously for a period of time, then ergosterol would begin to fluctuate.

Spectral Procedures

Spectral quantification of fungal contamination in forage and feed products began in 1960, as Birth (1960) estimated the smut content of wheat (*Triticum aestivum* L.) in the near infrared (NIR) region; the use of NIR to quantify fungal contamination continued in to the next decade as his group related levels of mold in corn (*Zea mays* L.) to reflectance, transmittance, and fluorescence from 400 to 800 nm (Birth & Johnson, 1970). To date, the most efficient procedure for quantification of mold in hay employs near infrared reflectance spectroscopy (NIRS), an empirical technology that is becoming increasingly applied to agriculture, industry, and medicine (Shenk et al., 1979). The NIRS procedure predicts concentrations of a given constituent from reflected light, with prediction equations developed by fitting reflectance data to chemical data (Martens & Naes, 1987). Much like the rationale that justifies remote sensing, NIRS technology is based on the coupled principles of spectroscopy and statistics. The NIRS procedure is very fast and precise, allowing a sample to be analyzed with excellent repeatability in 60 s; however, because NIRS is an empirical method, its accuracy and precision depend on the accuracy and precision of the chemical data. In addition, prediction equations can only be applied to populations whose samples are similar to those used to develop the equations. Recently developed software programs increase the possibility that populations for analysis match those used to develop NIRS equations (Shenk & Westerhaus, 1993).

The NIRS procedure has accurately quantified fungal chitin in alfalfa (*Medicago sativa* L.) and barley (*Hordeum vulgare* L.) (Roberts et al., 1987b, 1991). With both populations, NIRS-chitin was correlated to visual estimations of mold. In two nonreported experiments, an NIRS-mold equation was developed at the University of Illinois and applied to similar samples at the University of Kentucky (Dr. M. Collins, University of Kentucky, 1990, personal communication). In both studies, NIRS-chitin accurately predicted mold in alfalfa; in one of the studies, NIRS-chitin correlated well with visual estimations of mold when moisture failed to do so. In addition to these equations, other NIRS equations have quantified fungal spores in barley and wheat (Asher et al., 1982). The NIRS-spore equation also is very rapid, and regression statistics were better than those developed with chitin data. As mentioned above, however, the relationship between number of spores and amount of mycelia can vary.

It is difficult, if not impossible, to determine which functional groups are responsible for NIRS-mold quantification (Murray & Williams, 1987); however, there is substantial evidence that the functional groups detected are mycelial components. In one study, mycelia of *Penicillium* and *Aspergillus* were spiked into mold-free barley at naturally occurring levels, and the ensuing artificial calibration was successful (Roberts et al., 1991). In a similar study, mycelial dry matter in rice (*Oryza sativa* L.) was quantified by NIRS (Kojima, et al., 1994). Both studies independently reported that calibrations used wavelengths from 2348 to 2356 nm; this region corresponds to that reported for alfalfa mold calibrations (Roberts et al., 1987b).

MICROBIOLOGY OF SILAGES

When forages are preserved by ensiling, microorganisms such as lactic acid bacteria (LAB), clostridia, Enterobacteriaceae, and yeasts interact with the biochemical environment of the host tissue to affect both preservation and spoilage. The following sections will discuss the role of these microorganisms in the ensiling process, giving special attention to population dynamics and biochemical pathways.

Microorganisms Important to Ensiling

In his review, Beck (1978) discussed the history of identification, biochemical function, and early classification of silage microflora, including lactic acid bacteria, clostridia, Enterobacteriaceae, and yeasts.

Lactic acid bacteria are the most important beneficial microorganisms in the fermentation process. They are gram-positive, nonspore-forming bacteria. At present, LABs are classified on the basis of their biochemical abilities, products, and cell wall composition (Buchana & Gibbons, 1975). Although more than 2000 strains of lactic acid bacteria have been reported on corn silage, nearly all belong to subgenera of *Streptobacterium* and *Betabacterium*. They are most often grouped

Table 2-1. Common lactic acid bacteria in silage (from Edwards & McDonald, 1978).

Homofermentative	Heterofermentative
Lactobacillus plantarum	*Lactobacillus buchneri*
Pediococcus acidilactici	*Lactobacillus fermentum*
Streptococcus lactis	*Lactobacillus brevis*
Streptococcus faecium	*Lactobacillus viridescens*
Streptococcus faecalis	*Leuconostocs mesenteroides*

into two broad classes: homofermentative and heterofermentative, also called homolactic and heterolactic. Common species of each type are listed in Table 2–1.

Perhaps the most detrimental anaerobes are clostridia. Clostridia, also called butyric acid bacteria, are gram-positive, spore-forming bacteria that compete with lactic acid bacteria. According to Beck (1978), the two common types of clostridia include saccharolytic and proteolytic clostridia. Proteolytic clostridia, also called putrefying clostridia, are active in fermenting proteins, but not carbohydrates. Saccharolytic clostridia, however, primarily ferment carbohydrates. According to Beck, the most common clostridia in silage include *Clostridium tyrobutyricum, C. butyricum, C. sporogenes,* and *C. sphenoides.* Both saccharolytic and proteolytic clostridia have the ability to ferment lactic acid in the presence of acetate (Beck, 1978).

Enterobacteriaceae, also called coliform and acetic acid bacteria, are gram-negative, nonspore-forming anaerobes that convert carbohydrates primarily to acetic acid. They are most active during initial stages of fermentation. Because very little of their fermentation product is lactic acid, enterobacteriaceae are considered somewhat detrimental (Beck, 1978).

Microbial Activity Before Ensiling

Populations of microorganisms vital to the ensiling process undergo significant change long before forage enters the silo. Before forage is cut, LAB occur in very low levels; most are found in wounded or decaying plant material, and only ≈20% are homofermentative (Stirling & Whittenbury, 1963). After forage is cut, LAB increase up to 800-fold.

This post-cutting microbial growth is affected both by environment and management. Of the environmental effects, temperature and moisture during wilting most affect LAB growth. Muck (1989) reported that alfalfa wilting below 15°C was not conducive to LAB growth. He also reported that temperatures ranging from 22 to 25°C produced more LAB than temperatures of 19 to 22°C, and much more than 15 to 19°C. Muck also reported that LAB on alfalfa grew more at 40% DM than at 60%.

Of the management factors affecting LAB growth, wilting time and raking are very important. In the study cited above (Muck, 1989), LAB increased most rapidly during the first 48 h of wilting (Fig. 2–4). If forage was raked, the greatest increases occurred in the bottom of the swath (Table 2–2), while very little LAB grew on top of the swath. The authors hypothesized that failure of LAB to grow on top of the swath may be attributed to solar radiation or low moisture on the top of the swath. These factors, however, as well as rate of air flow on the top of the swath, were not investigated in this experiment.

Microbial Activity During the Ensiling Process

As summarized by Moser (Table 2–3), ensiling can be considered as a three-stage process. During the brief aerobic stage, the plant continues to respire, partially hydrolyzing cell walls and proteins, depleting O_2, and releasing CO_2. Dur-

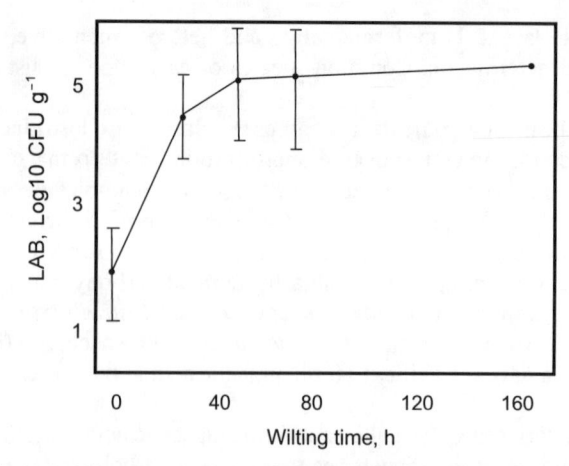

Fig. 2–4. Effect of wilting time on average lactic acid bacteria (LAB) counts from load samples where the average wilting temperature was between 20 and 25°C. Error bars denote one standard deviation. From Muck, 1989; CFU, colony forming units.

ing this stage, some of the sugars required by LAB for fermentation are metabolized, partly reducing the quality of silage.

Most of the O_2 is depleted after 1 d, and Stage 2 begins as LAB and Enterobacteriaceae grow rapidly. Coliform bacteria, yeasts, and molds decrease rapidly during the first week, then remain in low concentrations unless the anaerobic environment is compromised (Seale et al., 1981). This rate of bacterial growth in Stage 2 is affected by a moisture × pH interaction (Fig. 2–5). During this stage, LAB convert carbohydrates to lactic acid, acetic acid, ethanol, mannitol, acetaldehyde, and CO_2.

The preference for homofermentative over heterofermentative bacteria can be understood by reviewing fermentation reactions and corresponding energy requirements of the two types of bacteria (Table 2–4). As seen in metabolic pathways (Fig. 2–6, 2–7, and 2–8), homofermentative bacteria convert glucose and fructose to lactate with very few steps (Fig. 2–6), and their end products include a much higher proportion of lactic acid than those from heterofermentative bacteria (Fig. 2–7 and 2–8).

After about 3 wk, the pH of good quality silage decreases to the point of inhibiting further microbial growth (Fig. 2–9), and the silage is considered stable.

Table 2-2. Numbers of lactic acid bacteria on alfalfa prior to chopping (from Muck 1989).

Sample type	$\log_{10} CFU^{-1}$ g alfalfa
Fresh	--
After mowing	51
Before chopping	
Unraked, top of swath	223
Unraked, bottom of swath	1 266
Raked, top of swath	73
Raked, bottom of swath	18 683

Table 2-3. A summary of the stages of the ensiling process and corresponding bacteria and chemical changes (from Moser, 1980).

Time	Environment	Bacteria involved	Chemical changes
1 d	Stage I Anaerobic Stage; increase in temperature, depletion of O_2	Aerobic coliform and pigmented bacteria	Aerobic plant cell respiration; CO_2 production, CHO utilization, some protein hydrolyzed
1 d to 3 wk	Stage II Anaerobic Stage; fermentation, pH continuously decreasing	Lactic acid producing bacteria; depending on crop, many strains of bacteria can be isolated (e.g., *Lactobacillus*)	CHO converted to lactic, acetic, succinic acid; protein hydrolysis as amino acids; some hemicellulose breakdown; pH drops significantly
After 3 wk	Stage III Post-fermentation Storage Stage; should be relatively stable, have low pH and a pickled smell; may involve pH rise and spoilage if *Clostridium* or other sporeformers are active	Bacterial action stopped unless pH is not low enough, then *Clostridium* and other sporeformers are active	If bacterial action stopped, very little change, stable storage; if *Clostridium* are active, lactic acid converted to butyric and propionic acids; amino acids deaminated and decarboxylated forming ammonia and foul smelling amines

Table 2-4. Silage fermentation reactions and their associated losses (from McGechan, 1990).

Reaction	Microorganism	Dry matter loss	Energy loss
glucose → 2 lactic acid	homolactic	0	0.7
fructose → 2 lactic acid	homolactic	0	0.7
pentose → lactic acid + acetic acid	homolactic and heterolactic	0	1.7
glucose → 2 lactic acid + ethanol + CO_2	heterolactic	24.0	1.0
3 fructose + H_2O → lactic acid + 2 mannitol + acetic acid + CO_2	heterolactic	4.8	
2 fructose + glucose + H_2O → lactic acid + 2 mannitol + acetic acid + CO_2	heterolactic	4.8	
citrate + H_2O → 3 acetic acid + formic acid + CO_2	homolactic and heterolactic	14.0	
2 citrate → 2 acetic acid + acetoin + 4 CO_2	homolactic and heterolactic	46.0	
2 citrate + H_2O → 3 acetic acid + lactic acid + 3 CO_2	homolactic and heterolactic	29.7	+1.5
citrate + H_2 → acetic acid + ethanol + fermic acid + CO_2	homolactic and heterolactic		
malate → lactic acid + CO_2	homolactic and heterolactic	32.8	+1.8
2 malate → acetoin + 4 CO_2 + 4 H_2	homolactic and heterolactic	21.0	
malate + H_2O → acetic acid + formic acid + CO_2 + H_2	homolactic and heterolactic		
malate + 2 H_2 → ethanol + formic acid + CO_2	homolactic and heterolactic	58.0	
2 serine → acetoin + 2 CO_2 + 2 NH_3	homolactic and heterolactic	24.0	
arginine → ornithene + CO_2 + 2 NH_3			
glucose → 2 ethanol + 2 CO_2	yeast	48.9	0.2
fructose → 2 ethanol + 2 CO_2	yeast	48.9	0.2
glucose → butyric acid + 2 CO_2 + 2 H_2	clostridia, acting on WSC†		
2 lactic acid → butyric acid + 2 CO_2 + 2 H_2	clostridia, acting on WSC	51.1	18.4
3 alanine → 2 prophonic acid + acetic acid + 2 CO_2 + 3 NH_3	clostridia, acting on amino acids (oxidation/reduction)		
alanine + 2 glycine → 3 acetic acid + CO_2 + 3 NH_3	clostridia, acting on amino acids (oxidation/reduction)		
lysine → cadaverine + CO_2	clostridia, acting on amino acids (decarboxylation)		
valine → isobutyric acid + NH_3	clostridia, acting on amino acids (deamination)		
leucine → isovaleric acid + NH_3	clostridia, acting on amino acids (deamination)		

† WSC, water soluble CHO.

MICROBIOLOGY OF STORED FORAGES

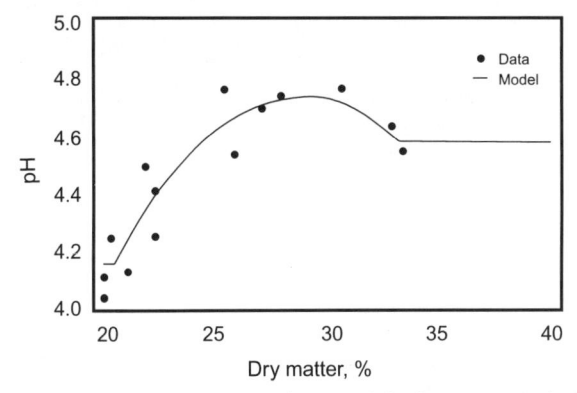

Fig. 2–5. Critical pH at which bacterial growth rate and death rate are equal, plotted against dry matter concentration. From Pitt et al., 1985.

The maximum pH required to prevent clostridial growth is a function of plant species and moisture (Seale et al., 1981; Fig. 2–10). A stable pH can be more difficult to achieve with legumes than with grasses because of legume buffering capacity (Moser, 1980); however, ease of maintaining a stable pH differs also among grass species (Seale et al., 1981).

If pH is not stabilized, clostridia will continue to be active. By redox, decarboxylation, and deamination, clostridia convert sugars, organic acids, and amino acids into spoilage products (Table 2–4). In addition to converting lactic acid to butyric acid, clostridia convert amino acids into unpleasant decay products such as cadaverine and putrescine (Edwards & McDonald, 1978).

When silage spoils after pH is stabilized, spoilage most often occurs by aerobic deterioration. This type of spoilage has been attributed primarily to yeasts

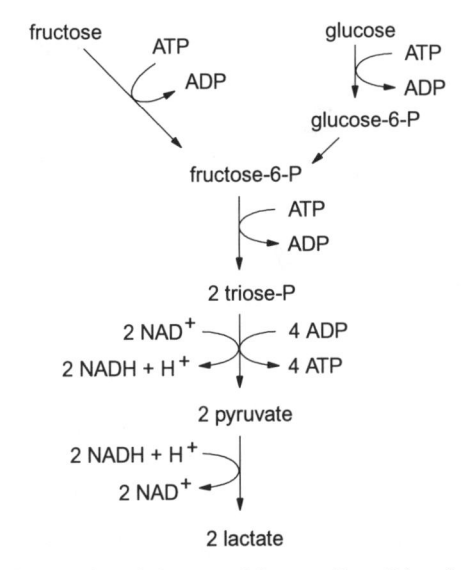

Fig. 2–6. Homolactic fermentation of glucose and fructose. From Edwards and McDonald, 1978.

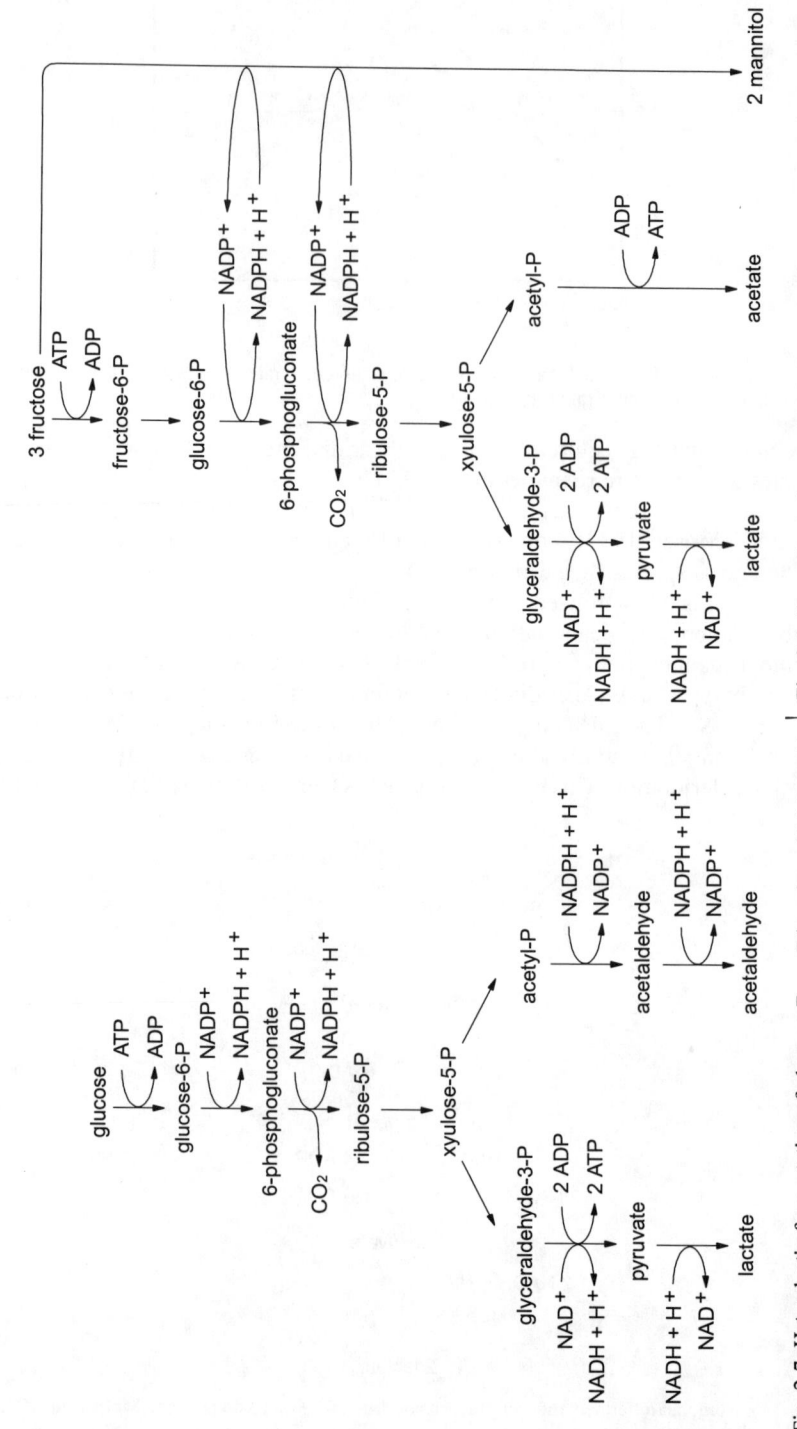

Fig. 2-7. Heterolactic fermentation of glucose. From Edwards and McDonald, 1978.

Fig. 2-8. Heterolactic fermentation of fructose. From Edwards and McDonald, 1978.

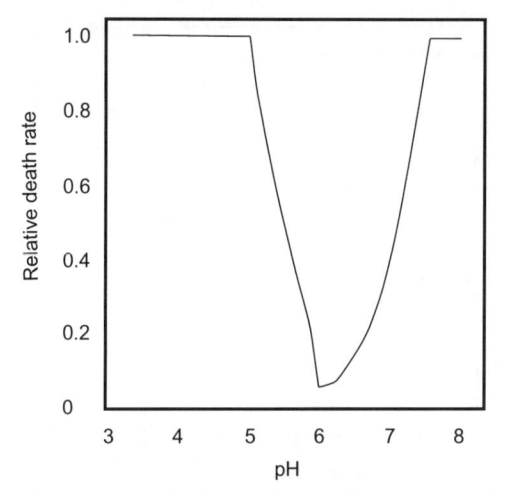

Fig. 2-9. Fraction of the maximum death rate of lactic acid bacteria at various pH levels. From Pitt et al., 1985.

and acetic acid bacteria (Courtin & Spoelstra, 1990). Courtin and Spoelstra presented a model based on the assumption that these organisms are primarily responsible for initial oxidation of lactic acid, acetic acid, and ethanol, reactions that facilitate pH increase, heat production, and CO_2 release. Their model accurately correlated these microbial and chemical trends (Fig. 2-11).

CONCLUSION

From a microbiological perspective, producing quality hay and silage is an exercise in microbe management. Forage producers must first be able to distinguish between beneficial and detrimental species. In order to regulate relative proportions of these species, producers also must understand which environmen-

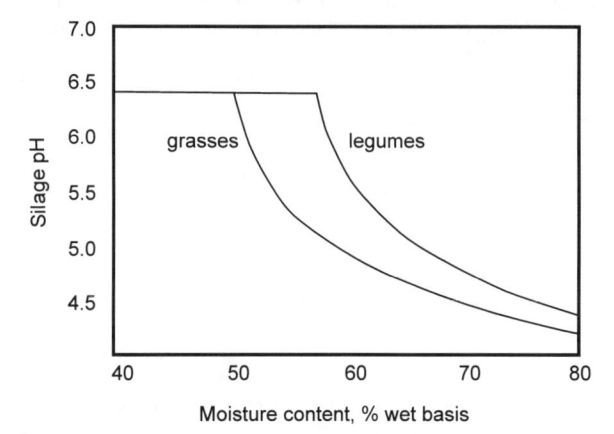

Fig. 2-10. The pH required to stop clostridial growth at varying moisture contents in grass and legume silages. From Leibensperger and Pitt, 1987.

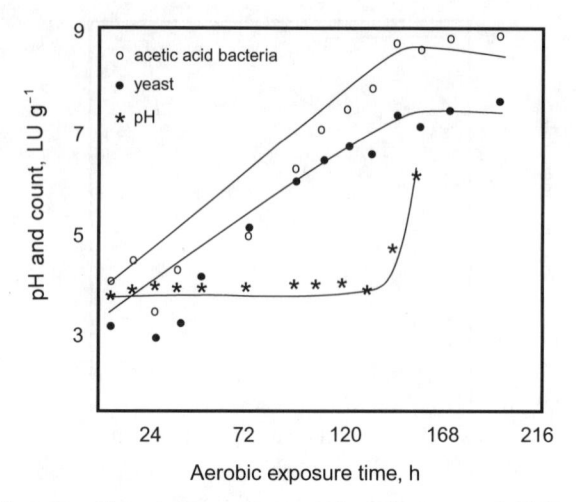

Fig. 2–11. Effect of aerobic exposure on acetic acid bacteria, yeast, and pH. Data predicted using model reported by Courtin and Spoelstra, 1990. LU, logarithmic units.

tal and managerial factors most often control population dynamics, factors such as plant species, moisture level, and O_2 status.

REFERENCES

Agarwal, V.K., and J.B. Sinclair. 1987. Detection of seedborne pathogens. p. 29–76. *In* Principles of seed pathology. Vol. 2. CRC Press, Boca Raton, FL.

Asher, M.J.C., I.A. Cowe, C.E. Thomas, and D.C. Cuthberston. 1982. A rapid method of counting spores of fungal pathogens by infrared reflectance analysis. Plant Pathol. 31:363–372.

Beck, T. 1978. The microbiology of silage fermentation. p. 63–115. *In* M.E. Micullough (ed.) Fermentation of silage: A review. National Feed Ingredients Assoc., Des Moines, IA.

Bethlenfalvay, G.J., R.S. Pacovsky, and M.S. Brown. 1981. Measurement of mycorrhizal infection in soybeans. Soil Sci. Soc. Am. J. 45:871–875.

Birth, G.S. 1960. Measuring the smut content of wheat. Trans. ASAE 3:19–21.

Birth, G.S., and R.M. Johnson. 1970. Detection of mold contamination in corn by optical measurements. J. Assoc. Am. Chem. 53:931–936.

Bishop, R.H., C.L. Duncan, G.M. Evancho, and H. Young. 1982. Estimation of fungal contamination in tomato products by a chemical assay for chitin. J. Food Sci. 47:437–444.

Buchana, R., and N. Gibbons. 1975. Bergey's manual of determinative bacteriology. Williams & Wilkins, Baltimore.

Cheng, A.P., and T.F. Boat. 1978. An improved method for the determination of galactosaminitol, glucosaminitol, glucosamine, and galactosamine on an amino acid analyzer. Anal. Biochem. 85:276.

Cherney, J.H., K.D. Johnson, J. Tuite, and J.J. Volenec. 1987. Microbial compositional changes in alfalfa hay treated with sodium diacetate and stored at different moisture contents. Anim. Feed Sci. Technol. 17:45–56.

Christensen, C.M., and H.H. Kaufmann. 1965. Deterioration of stored grains by fungi. Annu. Rev. Phyopathol. 3:69–84.

Cole, R.J., and R.H. Cox. 1981. Handbook of toxic fungal metabolites. Academic Press, New York.

Corbaz, R., P.H. Gregory, and M.E. Lacey. 1963. Thermophilic and mesophilic actinomycetes in mouldy hay. J. Gen. Microbiol. 32:449–455.

Courtin, M.G., and S.F. Spoelstra. 1990. A simulation of the microbiological and chemical changes accompanying the initial stage of aerobic deterioration of silage. Grass Forage Sci. 45:153–165.

Cousin, M.A., C.S. Ziedler, and P.E. Nelson. 1984. Chemical detection of mold in processed foods. J. Food Sci. 49:439–445

Edwards, R.A., and P. McDonald. 1978. The chemistry of silage fermentation. p. 29–60. *In* M.E. Micullough (ed.) Fermentation of silage: A review. Natl. Feed Ingredients Assoc., Des Moines, IA.

Goering, H.K., and C.H. Gordon. 1973. Chemical aids to prevention of high moisture feeds. J. Dairy Sci. 56:1347–1351.

Gregory, P.H., M.E. Lacey, G.N. Festenstein, and F.A. Skinner. 1963. Microbial and biochemical changes during the molding of hay. J. Gen. Microbiol. 33:147–174.

Golubchuk, M., L.S. Cuendet, and W.F. Geddes. 1960. Grain storage studies: XXX. Chitin content of wheat as an index of mold contamination and wheat deterioration. Cereal Chem. 37:405–411.

Hintz, H.F., and J.E. Lowe. 1977. Health of the horse. p. 555–572. *In* J.E. Warren (ed.) The horse. W.H. Freeman & Co., New York.

Howard, B.J. 1911. Tomato ketchup under the microscope with practical suggestions to insure a clean product. USDA Bull. Chem. Circ. 68. USDA, Washington, DC.

Hubbard, J.D., L.M. Seitz, and H.E. Mohr. 1979. Determination of hexosamines in chitin by ion-exchange chromatography. J. Food Sci. 44:1552–1553.

Jarvis, B. 1977. A chemical method for the estimation of mould in tomato products. J. Food Technol. 12:581–591.

Jaster, E.H. 1995. Legume and grass silage preservation. p. 91–115. *In* K. Moore and M. A. Peterson (ed.) Post-harvest physiology and preservation of forages. CSSA Spec. Publ. 22. CSSA and ASA, Madison, WI.

Jeffers, D.L., A.F. Schmitthenner, and D.L. Reichard. 1982. Seed-borne fungi, quality, and yield of soybeans treated with benomyl fungicide by various application methods. Agron. J. 74:589–592.

Jones, A.R., L.M. Bezeau, B.D. Owen, and F. Whiting. 1955. The effects of mold growth on the digestibility and feeding value of grains for swine and sheep. Can. J. Agric. Sci. 35:525–532.

Jones, A.L., R.E. Morrow, W.G. Hires, G.B. Garner, and J.E. Williams. 1985. Quality evaluation of large round bales treated with sodium diacetate or anhydrous ammonia. Trans. ASAE 28:1043–1046.

Kaspersson, A., R. Hlodversson, U. Palmgren, and S. Lindgren. 1984. Microbial and biochemical changes occurring during deterioration of hay and preservative effect of urea. Swed. J. Agric. Res. 14:127–133.

Kojima, Y, U. Asai, Y. Hata, E. Ichikawa, A. Kawato, and S. Imayasu. 1994. Estimation of mycelial weight in rice *Koji* by near infrared reflectance spectroscopy. Nippon N gei Kagaku Kaishi 68:801–807.

Lacey, J., and J. Dutkiewicz. 1976. Methods for examining the microflora of moldy hay. J. Appl. Bacteriol. 41:13–27.

Lacey, J., and K.A. Lord. 1977. Methods for testing chemical additives to prevent molding of hay. Annu. Appl. Biol. 87:327–335.

Lacey, J., K.A. Lord, and G.R. Cayley. 1981. Chemicals for preventing molding in damp hay. Anim. Feed Sci. Technol. 6:323–336.

Lacey, J., K.A. Lord, H.G.C. King, and R. Manlove. 1978. Preservation of baled hay with propionic and formic acids and a proprietary additive. Annu. Appl. Biol. 88:65–73.

Leibensperger, R.Y., and R.E. Pitt. 1987. A model of clostridial dominance in ensilage. Grass Forage Sci. 42:297–317.

Lin, H.H., and M.A. Cousin. 1985. Detection of mold in processed foods by high performance liquid chromatography. J. Food. Prot. 8:671–678.

Magan, N., and J. Lacey. 1987. The influence of water and temperature on the growth of fungi causing spoilage of stored products. p. 43–52. *In* T.J. Lawson (ed.) Proc. of Stored Products Pest Control Symp., Univ. of Reading, Berkshire. 25–29 Mar. 1987. Univ. of Reading, Reading, England.

Martens, H., and T. Naes. 1987. Multivariate calibration by data compression. p. 57–87. *In* P.C. Williams and K. Norris (ed.) Near-infrared technology in the agricultural and food industries. Am. Assoc. Cereal Chem., St. Paul, MN.

Mawhinney, T.P. 1986. Simultaneous determination of N-acetylglucosamine, N-acetyl galactosamine, N-acetylglucosaminitol, and N-acetylgalactosaminitol by gas-liquid chromatography. J. Chromatogr. 351:91–102.

Mawhinney, T.P., M.S. Feather, G.J. Barbero, and J.R. Martinez. 1980. The rapid, quantitative determination of neutral sugars (as aldononitrile acetates) and amino sugars (as O-methyloxime acetates) in glycoproteins by gas-liquid chromatography. Anal. Biochem. 101:112–117.

McGrechan, M.B. 1990. A review of losses arising during conservation of grass forage: 2. Storage losses. J. Agric. Eng. Res. 45:1–30.

Moser, L.E. 1980. Quality of forage as affected by post-harvest storage and processing. p. 227–260. *In* C.S. Hoveland (ed.) Crop quality, storage, and utilization. ASA and CSSA, Madison, WI.

Muck, R.E. 1989. Initial bacterial numbers on lucerne prior to ensiling. Grass Forage Sci. 44:19–25.

Murray, I., and P.C. Williams. 1987. Chemical principles of near-infrared technology. p. 17–34. *In* P.C. Williams and K. Norris (ed.) Near-infrared technology in the agricultural and food industries. Am. Assoc. Cereal Chem., St. Paul, MN.

Pitt, R.E., R.E. Muck, and R.Y. Leibensperger. 1985. A quantitative model of the ensilage process in lactate silages. Grass Forage Sci. 40:279–303.

Plassard, C.S., D.G. Mousain, and L.E. Salsac. 1982. Estimation of mycelial growth of basidiomycetes by means of chitin determination. Phytochemistry 21:345–348.

Ride, J.P., and R.B. Drysdale. 1972. A rapid method for the chemical estimation of filamentous fungi in plant tissue. Physiol. Plant Pathol. 2:7–15.

Roberts, C.A., R.R. Marquardt, A.A. Frohlich, R.L. McGraw, R.G. Rotter, and J.C. Henning. 1991. Chemical and spectral quantification of mold in barley. Cereal Chem. 68:272–275.

Roberts, C.A., K.J. Moore, D.W. Graffis, H.W. Kirby, and R.P. Walgenbach. 1987a. Chitin as an estimate of mold in hay. Crop Sci. 27:783–785.

Roberts, C.A., K.J. Moore, D.W. Graffis, H.W. Kirby, and R.P. Walgenbach. 1987b. Quantification of mold in hay using near infrared reflectance spectroscopy. J. Dairy Sci. 70:2560–2564.

Seale, D.R., C.M. Quinn, and P.A. Whittaker. 1981. Microbiological and chemical changes during the first 22 days of ensilage of different grasses. Isr. J. Agric. Res. 20:61–70.

Seitz, L.M., and Y. Pomeranz. 1983. Ergosterol, Ergosta-4,6,8(14),22-tetraen-3-one, ergosterol peroxide, and chitin in ergoty barley, rye, and other grasses. J. Agric. Food Chem. 31:1036–1038.

Seitz, L.M., D.B. Sauer, R. Burroghs, H.E. Mohr, and J.D. Hubbard. 1979. Ergosterol as a measure of fungal growth. Phytopathology 69:1202–1203.

Shenk, J.S., and M.O. Westerhaus. 1993. Using near infrared reflectance product library files to improve prediction accuracy and reduce calibration costs. Crop Sci. 33:578–581.

Shenk, J.S., M.O. Westerhaus, and M.R. Hoover. 1979. Analysis of forages by infrared reflectance. J. Dairy Sci. 62:807–812.

Stirling, A.C., and R. Whittenbury. 1963. Sources of the lactic acid bacteria occurring in silage. J. Appl. Bacteriol. 26:86–90.

Swift, M.J. 1973. The estimation of mycelial biomass by determination of the hexosamine content of wood tissue decayed by fungi. Soil Biol. Biochem. 5:321–332.

Tsuji, A., T. Kinoshita, and M. Hoshina. 1969. Microdetermination of hexosamines. Chem. Pharm. Bull. 17:1505–1510.

Weete, J.D. 1974. Fungal lipid biochemistry: Distribution and metabolism. Plenum Press, New York.

Wessels, J.G.H., and J.H. Sietsma. 1981. Fungal cell walls: A survey. p. 352–394. *In* W. Tanner and F. A. Loewus (ed.) Plant carbohydrates: II. Extracellular carbohydrates. Encyclopedia of Plant Physiology. Vol. 13B. Springer-Verlag, Berlin.

Wittenberg, K.M., and S.A. Moshtaghi-nia. 1991. Influence of anhydrous ammonia and bacterial preparations on alfalfa forage baled at various moisture levels: II. Fungal invasion during storage. Anim. Feed Sci. Technol. 34:67–74.

Wittenberg, K.M., S.A. Moshtaghi-nia, P.A. Mills, and R.G. Platford. 1989. Chitin analysis of hay as a means of determining fungal invasion during storage. Anim. Feed Sci. Technol. 27:101–110.

Woolford, M.K. 1984. The antimicrobial specta of organic compounds with respect to their potential as hay preservatives. Grass Forage Sci. 39:75–79.

Woolford, M.K., and R.M Tetlow. 1984. The effect of anhydrous ammonia and moisture content on the preservation and chemical composition of perennial ryegrass hay. Anim. Feed Sci. Technol. 11:159–166.

3 Field Curing of Forages

C. Alan Rotz

USDA Agricultural Research Service
East Lansing, Michigan

Field cured forages have an important role in feeding ruminant animals in the temperate climates of the world. Forage growth is not possible for much of the year in these areas. Large amounts of forage must be mechanically harvested and stored for use when forage cannot be grown. Although grazing can be used during summer months, the trend during the past 30 yr has been to use less grazing and more conserved forage. Despite a recent shift toward more grazing, most ruminant animals in these climates are fed largely or entirely from conserved feeds.

Field curing is required to conserve most forage crops. Field curing is a process that uses the field environment to remove moisture from the crop to obtain a moisture level suitable for storage. The crop is mowed and laid on the field surface to wilt or dry. Growing forage typically contains 700 to 850 g kg^{-1} of moisture (wet basis), i.e. 2.3 to 5.6 parts water for each part of plant dry matter. Plant material with this much moisture is difficult and expensive to preserve for long storage periods. Even when sealed in a silo, improper fermentation causes a high loss of plant dry matter and a degradation of the protein and other nutrients required by animals. Moisture removal is necessary to obtain stable preservation of forage as dry hay or wilted silage.

Field curing involves both drying and rewetting processes. As the crop lays in the field, moisture moves between the crop and its environment until a suitable crop moisture content is attained for harvest. This moisture movement cycles between moisture loss and moisture gain in the crop as the environment changes relative to the crop. Unless rain occurs, daytime periods predominantly produce moisture loss or drying. At night, high humidities, dew, and perhaps rainfall produce a gain in crop moisture.

Many factors affect the rate and extent of drying during field curing. Solar radiation, air temperature, humidity, soil moisture, and other environmental factors influence both the drying and rewetting processes. Crop yield, stem diameter, leaf to stem ratio, swath structure, and other crop characteristics also can enhance or impede moisture movement. With sunny days, warm temperatures, dry soil, and a thin swath, relatively fast curing is attained. Less favorable conditions and periodic rainfall prolongs the curing process. Forage that has laid in the

---

Copyright © 1995 Crop Science Society of Agronomy and American Society of Agronomy, 677 S. Segoe Rd., Madison, WI 53711, USA. *Post-Harvest Physiology and Preservation of Forages.* CSSA Special Publication no. 22.

field for two or more weeks is often not suitable as animal feed and must be destroyed.

Traditionally, hay harvest is the most common method for conserving grass and legume forage crops. Forage dried to contain 150 to 180 g kg^{-1} moisture is baled and stored up to a year and even more with relatively little loss or change in the nutrients available as animal feed. Typical field curing times to produce hay range from 2 to 7 d. Less than 5 d are required when forage is spread in thin swaths and dried with relatively good drying conditions. In drier climates, forage is often dried more slowly in thick swaths or windrows.

Field curing of hay requires the removal of about 3 t of moisture for each tonne of hay produced. Each kilogram of moisture removed requires ≈2.3 MJ of energy to transform it from a liquid to a vapor. This means that each tonne of hay produced requires 7 billion Joules of energy to evaporate the moisture. This is approximately the amount of energy available from 270 L of fuel oil. For field cured forage, this energy is supplied by the sun. Drying with other forms of energy is generally impractical and uneconomical.

Wilted silage is another popular method for conserving forage, particularly on dairy farms. Forage ensiles best when dried to a moisture content between 500 and 650 g kg^{-1} (wet basis). This requires the removal of 2 to 3 t of moisture for each tonne of forage dry matter produced. Under good drying conditions, this wilting requires 1 to 3 d. With less time laying in the field, the crop is less likely to receive rain damage and the resulting nutrient losses.

Although field curing requires the removal of a large amount of moisture, this amount is less than that transpired by the crop on an average day. Considering that the evapotranspiration rate is normally greater than 5 cm d^{-1} (Peterson, 1972), a crop yielding 5 t dry matter (DM) ha^{-1} transpires more than 10 t of moisture per tonne of crop DM each day. Field curing, therefore is relatively inefficient compared with the natural evapotranspiration of the crop.

To better understand the field curing process and methods for enhancing drying, some knowledge of the basics of moisture movement is useful. This chapter reviews a few basic principles of moisture movement as they apply to the drying and rewetting of forage. Using these principles, the many crop and environmental factors that affect moisture movement are discussed in detail. With this understanding of field curing, equipment, and procedures for reducing the field curing time are discussed.

PHYSICAL PRINCIPLES OF MOISTURE MOVEMENT

The primary goal in field curing is to remove moisture from the crop. Other volatiles such as alcohol and fats also may be removed, but the quantity lost is very small relative to moisture (Weeks & Whitney, 1964). Three major processes of moisture movement are involved. Evaporation is the transformation of moisture from a liquid to a vapor (or gas) that can be contained in the surrounding air. The reverse of evaporation is condensation where vapor transforms to a liquid. The other process, diffusion, is the migration or movement of vapor from a higher level of concentration to a lower level of concentration.

Moisture Movement in the Plant

Field curing begins with the mowing of a living plant. An important process in living plants is the removal of moisture by evaporation. This process known as transpiration may be regarded as a continuous flow of water by potential gradients from the groundwater level through the soil and plant roots to the leaves, where it is transformed by solar energy into water vapor (Rijtema, 1968). Moisture moves through the xylem and parenchyma in the core of the stem to the leaf. Solar radiation provides energy for transpiration. When the radiant energy is incident on plant leaves, moisture is evaporated and the vapor diffuses into the surrounding air. Loss of moisture from the leaves causes a moisture gradient within the plant that draws more moisture through the plant from the soil.

The surface of forage plants is covered by a protective layer of cells called the epidermis. The outer surface of the epidermis is a waxy cuticle that is relatively impervious to moisture. Functions of this covering include the prevention of physical abrasion of plant tissue, leaching of plant components, and excessive moisture loss from the growing plant. Most moisture transpired by the plant must exit through stomata. These small holes in the epidermis cover only 1 to 3% of the plant surface, but 80 to 90% of the moisture leaving the plant passes through the stomata (Sullivan, 1973). Legumes generally contain more stomata than grasses. Stomatal numbers vary with cultivar, growing conditions and plant age, but the ratio of pore area to leaf area appears to be relatively consistent (Pedersen & Buchele, 1960a; Harris & Tulberg, 1980).

Stomata are the plant's mechanism for controlling moisture loss. When open, moisture freely evaporates from the plant surface drawing more moisture through the plant. Stomata can close to some extent without reducing transpiration, but when completely closed, moisture loss is greatly reduced to that which migrates through the epidermis (Harris & Tulberg, 1980). Stomata respond to light, temperature, water availability, and other stimuli. They generally close at night and open during the day (Whitney et al., 1969). Insufficient water availability may lead to partial closure during the middle of the day with complete closure under more severe moisture stress (Pedersen & Buchele, 1960a).

When the forage crop is cut and laid on the field surface to dry, moisture loss initially continues much as in living plants. Since the stem and leaves are severed from the root, lost moisture is no longer replaced and wilting begins. Immediately after cutting, stomata openings may increase, but they soon decrease as drying continues. Stomatal closure is complete for most species within a couple hours after cutting and this occurs before a third of the initial water is lost (Jones & Harris, 1979; Jones & Palmer, 1932; Harris & Tulberg, 1980). As stomata close, the moisture loss or drying rate decreases. Treatments used to hold stomata open during drying have significantly increased the drying rate of alfalfa (*Medicago sativa* L.) leaves (Whitney et al., 1969).

As drying continues, moisture moves both axially along the stem and radially toward the surface of the stem. The natural passage of moisture along the stem and through the leaf is the primary pathway in the early stages of drying. At least 35% of the moisture contained in the alfalfa stem at cutting exits the plant through the leaf during field drying (Harris & Tulberg, 1980). For this reason, the drying rate of intact forage plants is considerably greater than that of equivalent amounts of detached leaves and stems (Harris & Tulberg, 1980; Jones, 1979; Jones & Harris, 1979). When the plant dries to a moisture content of ≈400 g kg^{-1}, dried tissue and air spaces develop in the xylem and parenchyma. These pockets break the moisture gradient within the plant impeding or eliminating further moisture movement through this pathway.

As axial movement of moisture slows, radial migration to the stem surface becomes the predominant pathway for further moisture loss. The drying rate continues to decrease due to the greater resistance to moisture movement in the radial direction. The rate at which moisture moves through the plant is modeled as diffusivity. The radial diffusivity of moisture in alfalfa stems is a factor of 10 less than the axial diffusivity. The epidermal layer is the primary restraint with a diffusivity 1000 times less than the radial diffusivity (Bagnall et al., 1970). Removing the epidermis can greatly increase the drying rate of forages with the greatest effect on leaf petioles and stems at low moisture contents (Bagnall et al., 1970; Harris & Shanmugalingam, 1982).

Moisture Movement to the Environment

After migrating to the surface of the plant, moisture must evaporate to leave the plant. Once transformed to a gaseous state (vapor), the moisture moves away from the plant following the principles of moisture diffusion. Diffusion is controlled by a gradient in vapor pressure. Air at the plant surface is nearly saturated (100% relative humidity) holding as many water molecules as it can at that temperature. The saturated vapor pressure of this air increases exponentially with temperature, i.e., small increases in temperature yield greater increases in vapor pressure. In the ambient air outside the swath, vapor pressure is independent of temperature. The vapor pressure of the ambient air increases with relative humidity toward the saturated state at a relative humidity of 100%.

The difference in vapor pressure between the plant surface and the ambient air, therefore, is primarily a function of temperature and secondarily a function of humidity. As temperature is increased, the saturated vapor pressure rapidly increases, while the vapor pressure in the ambient air remains constant unless there is some change in humidity. The difference between the saturated and ambient vapor pressure is called vapor pressure deficit. Moisture moves from the relatively high vapor pressure at the plant surface toward the ambient air, and the rate of drying increases with increased vapor pressure deficit.

When the relative humidity of the ambient air is very high, such as often occurs at night, the vapor pressure gradient may reverse, drawing moisture back into the plant. If the temperature of the plant drops below the temperature of the saturated air (dew point), moisture condenses on the surface of the plant where some is absorbed into the plant. Thus, dew formation and rewetting occur.

The rate at which moisture moves to and from the curing forage is controlled by two physical restraints. The first is a boundary layer of moist air that surrounds the plant. Air molecules at the plant surface have little movement. Molecular movement increases as the distance from the plant surface increases until it is the same as that of the surrounding air. This region of transition between the plant surface and ambient air is the boundary layer.

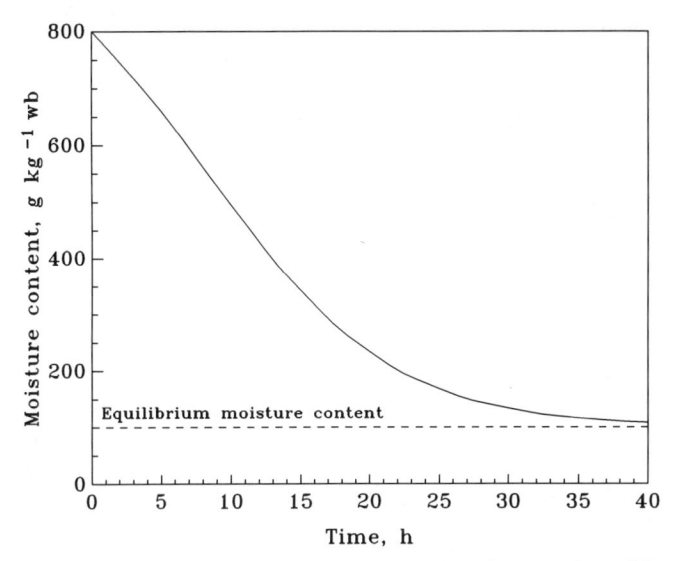

Fig. 3–1. Typical drying curve for forage dried under fixed environmental conditions; wb, wet basis.

The resistance of the boundary layer to moisture movement is directly related to the thickness of the layer. As air movement near the plant is increased, the thickness of the boundary layer decreases and drying improves. During field curing, temperature gradients in the air near the plant cause convective air currents. These convective currents are greatest when incident solar energy is adding heat to the plant as well as the air and soil near the plant. Wind also may cause air movement near the plant.

The second physical restraint to drying is created by the swath structure. As the thickness and density of the swath increase, movement of air within the swath becomes more restricted. The relatively still air inside the swath forms a microclimate with a relative humidity near 80%. This moist air reduces the gradient in vapor pressure between the plant and surrounding air and thus slows drying. The still air inside the swath also increases the thickness of the boundary layer surrounding the plant, which further slows drying.

The moisture content of a thin layer of forage plants dried in a constant environment follows a curve where the rate of moisture loss decreases as the plants dry (Fig. 3–1). Early in the drying process the drying rate (loss of moisture per unit of plant dry matter) is very high. Fast drying is possible because the plants initially contain abundant moisture and that moisture is near the plant surface where it can readily transpire. As the plants dry, the drying rate steadily decreases to zero at an equilibrium moisture content where the vapor pressure for moisture contained in the plant is the same as that in the surrounding environment. The plants can remain in this environment for an indefinite time with no further change in moisture content.

Forage dries differently in the field compared with the laboratory. Continuous change in the environment surrounding the swath causes large fluctuations in the drying rate. A typical drying curve for field cured forage has several

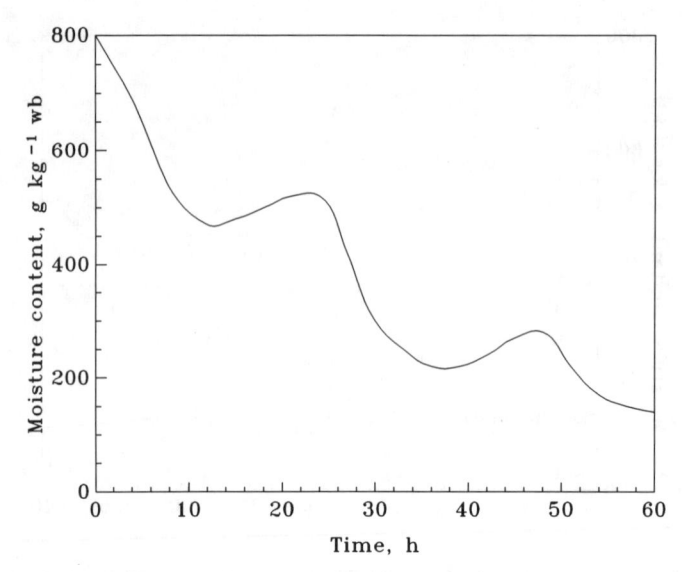

Fig. 3–2. Typical drying curve for forage dried in the field with good weather conditions; wb, wet basis.

rewetting and drying cycles before the crop approaches a moisture content suitable for harvest (Fig. 3–2). Drying rate expressed as moisture lost per unit of moisture contained in the forage is nearly independent of the amount of moisture contained in the crop throughout the moisture range of interest in field curing (Rotz & Chen, 1985). This measure of drying rate, therefore, is primarily influenced by the characteristics of the plant, swath, and drying environment. Measurement of the average rate of moisture loss per unit of available crop moisture provides a good tool for comparing differences in forage treatments dried under the same environmental conditions (Rotz & Sprott, 1984).

Constraints to Moisture Movement

The field drying rate of forage is limited by characteristics of the forage plants, swath, and the surrounding environment. Most of the time all three of these factors have some effect on limiting drying; however, the predominant constraint varies from one set of conditions to another. When the environmental conditions are not conducive to good drying, little can be done to the plant or swath to improve drying. More specifically, if the humidity is high, solar radiation is low and air temperature is low, the vapor pressure gradient is low. With a small difference in vapor pressure between the plant surface and ambient air, little moisture movement occurs. Crushing the plant, spreading the swath, or any other treatment to the plant or swath has little effect on drying because the environment is the predominant constraint.

Swath structure also may restrict drying. When forage is laid in a thick, dense swath, the resistance created by the swath can become the principal constraint to drying. Alteration of the plant has little effect on drying since the moisture cannot readily move out of the swath. Even when the environment is ideal for drying, the drying rate may be relatively low due to the constraint of the swath. When forage is dried in a thin open swath under good environmental drying conditions, the plant becomes the predominant constraint to moisture movement. Crimping, crushing, or other alteration of the plant provides the greatest increase in drying rate under this scenario.

Most field curing occurs with each of these constraints placing some control on drying. The relative impact each of the major constraints has on drying varies throughout the day (Clark & McDonald, 1977). Early and late in the day, the environment normally provides the greatest constraint to drying. During much of a sunny day, the drying rate is limited more by swath and plant constraints. The effect each has on drying changes continuously throughout the curing process as characteristics of the plant, swath, and environment change.

FACTORS INFLUENCING THE DRYING OF FORAGE

Many plant, swath, and environmental factors affect the constraint each of these places on the rate and extent of forage drying. Drying rates vary from zero or no moisture loss to very high levels where the crop may lose 50% of its contained moisture each hour. Under certain conditions the crop dries to a point where little moisture is left in the plant, but at other times it may lay in the field for days with a moisture content of 300 to 400 g kg^{-1}. The rate of drying is controlled by many crop and environmental factors, and the extent of drying is limited to the equilibrium moisture content. Equilibrium moisture is primarily related to weather variables, but crop characteristics also have some influence.

Crop Factors

Crop factors that influence drying include the initial moisture content of the crop, the forage species and physical characteristics of the forage. Forage moisture content at the time of mowing is normally between 700 and 850 g kg^{-1} (wet basis). Forage with 850 g kg^{-1} of moisture contains more than twice as much water as forage at 700 g kg^{-1} moisture. Water contained at a high moisture content exits the plant more quickly than that at a lower moisture content, but the added moisture still increases field curing time up to half a day.

Initial crop moisture varies with crop maturity, environmental conditions, and time of day. Crop maturity has the greatest effect. In the early stages of development, most forage plants contain 800 to 850 g kg^{-1} moisture. As the crop matures to a reproductive stage, the moisture content drops as low as 600 g kg^{-1}. At a desirable stage of maturity for harvest, most forages contain 750 to 800 g kg^{-1} moisture. Forage grown in dry conditions normally contains slightly less moisture than that grown in more humid conditions with ample soil moisture. Moisture content also may vary during the day with the highest moisture in the early morning. More moisture is transpired than can be taken up through the roots causing up to a 30 g kg^{-1} decrease in moisture content by late afternoon (Harris & Tulberg, 1980). Plant species probably influences initial moisture; however, the interaction with crop maturity and environmental factors makes it difficult to distinguish consistent differences among species.

Species effects on drying rates are documented. In general, grass species dry faster than legume species. Among legumes, red clover (*Trifolium pratense*

L.) is generally regarded as slower drying than alfalfa. When dried under the same conditions in a laboratory, birdsfoot trefoil (*Lotus corniculatus* L.) dried a little slower than alfalfa, but both dried faster than red clover (Johnson et al., 1984; Clark et al., 1985). Among grass species, tall fescue (*Festuca arundinacea* Schreber) is known for rapid moisture loss, drying up to four times faster than perennial ryegrass (*Lolium perenne* L.; Jones & Harris, 1979). Drying rates of Italian ryegrass (*L. multiflorum* Lam.), meadow fescue (*F. pratensis* Hudson), timothy (*Phleum pratense* L.), and cocksfoot (*Dactylis glomerata* L.) fall between those of tall fescue and perennial or hybrid ryegrass (Owen & Wilman, 1983). Coastal bermudagrass [*Cynodon dactylon* (L.) Pers.] dries rather slowly with drying rates less than that of alfalfa (Person & Sorenson, 1970).

Drying rate differences among species can be attributed to differences in the physical characteristics of plants. Species with the highest surface-area to dry-weight ratios generally provide the fastest drying (Morris, 1972). Thus, leaves dry more quickly than stems. Leaves also are the natural pathway for moisture loss from plants. When all other conditions are similar, a leafy forage dries more quickly than forage with fewer leaves (Clark et al., 1985; Owen & Wilman, 1983). Leaf to stem ratio varies among species, but it also varies among cultivars, growing conditions, and harvest date. First-cutting alfalfa normally has a lower leaf to stem ratio than later cuttings. Also, alfalfa grown in high elevations or under mild drought stress may have a higher leaf to stem ratio.

Drying rate of forage crops tends to decrease with increased stem diameter (Clark et al., 1985). In the later stages of drying, moisture in the core of the stem becomes difficult to remove. The natural flow of moisture along the stem is blocked, so the moisture must move radially from the stem. Therefore, an increase in the radial distance to the surface tissue (stem diameter) slows drying. Stem diameter also varies among species, cultivars, growing conditions, and harvest dates. Thicker stem material is one of several factors that contribute to slower drying of first cutting alfalfa grown in humid northern climates (Rotz et al., 1987).

Physical properties of the surface of the forage plant may also influence drying. Epicuticular wax content varies as much as 300% among alfalfa cultivars and clones (Galeano et al., 1986). Although not demonstrated, it is logical that forage with a thinner cuticle will dry more quickly when swath and environmental constraints are not limiting drying. Growing conditions also can affect epicuticular wax formation. Field grown alfalfa has a lower drying rate than greenhouse-grown alfalfa when both are of similar maturity (Johnson et al., 1984). The difference is attributed to the influence of temperature, relative humidity and photoperiod on epicuticular wax formation.

Another physical characteristic of the plant surface is the trichome or pubescence density. Trichomes are small hair-like structures on the plant surface that are most commonly found on red clover. Drying rate decreases with an increase in trichome density (Collins et al., 1991). The presence of trichomes reduces air movement near the plant surface, i.e., they increase the thickness of the boundary layer surrounding the forage plant. A thicker boundary layer restricts moisture loss, particularly when all other swath and environmental restrictions are reduced or removed.

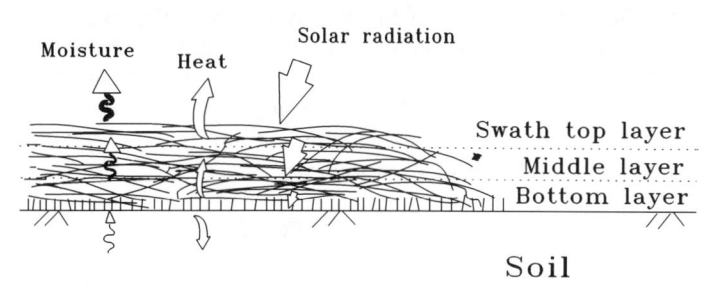

Fig. 3–3. Solar radiation, heat and moisture movement through layers of a swath during field curing.

Plant maturity also influences drying rate. In the laboratory, grass at a vegetative stage of maturity dries faster then that in a reproductive stage (Jones, 1979; Menzies & O'Callaghan, 1971). Also, younger tissue of the apical internode of alfalfa dries more quickly than the second and third internodes (Clark et al., 1985). Greater drying rates in the less mature plant tissue is due to greater initial moisture contents, greater leaf to stem ratios, thinner stems and perhaps differences in the cuticle. With field drying, however, immature grass may dry slower than grass in a reproductive stage of development (Savoie & Mailhot, 1986; Savoie et al., 1984). The less mature grass contains more moisture and it collapses into a more dense swath that restricts drying.

Swath thickness and density are closely related swath characteristics that often influence drying. Thickness is the distance between the top and bottom surfaces of the swath. This distance is controlled by the width of the swath laid on the field surface and the crop yield. Narrower swaths and heavier yields cause thicker swaths. Density is the amount of crop material per unit volume of the swath and thicker swaths tend to be more dense. Drying rate decreases as thickness and density increase.

A swath of field curing forage can be viewed as several layers of plant material (Bruck & van Elderen, 1969; Clark & McDonald, 1977). The upper layer exposed to solar radiation and ambient air dries most quickly (Fig. 3–3). Much of the solar radiation is absorbed or reflected by the upper layer so less is available in lower layers to enhance drying. Each layer also creates an insulating barrier for heat and moisture transfer to lower layers. A thick, dense swath restricts convective air currents within the swath that carry moisture from the bottom layers to the swath surface. Restricted movement creates a humid microclimate within the swath. Low solar radiation, high humidity, and poor air movement all lead to slow drying of lower layers of the swath.

As drying progresses, the moisture content of the surface layer drops to a threshold level where the plant resistance to moisture loss limits evaporation (Clark & McDonald, 1977). The moisture content of the second layer then decreases to approach the threshold moisture level. The lowest layer dries slowly until plant resistance becomes limiting. The swath moisture content at any stage of wilting is the mean of the widely varying moisture contents of all layers.

Spreading the swath creates thinner layers within the swath and exposes more crop to the ambient environment on the upper layer. Turning or inverting of the swath can speed drying by exposing the more moist bottom of the swath to the more favorable drying conditions on the upper surface. When a less dense, fluffy swath or windrow is maintained, more air movement occurs through the swath reducing the stratification of moisture.

Environmental Factors

The ambient weather and soil form the environment for the field curing swath. Weather variables have the greatest effect, but soil properties also influence drying. Major weather variables are solar radiation, air temperature, air humidity, and wind velocity. A high correlation among variables makes it difficult to distinguish between the effects each has on drying. Both solar radiation and temperature begin at a minimum level at dawn, increase to a maximum level around mid-day, and decrease toward dusk. The diurnal cycle of air temperature normally lags solar level by a few hours and continues a slow decrease throughout the night. Relative humidity of the ambient air is high at night, but decreases during the day as temperature increases. Wind velocity is somewhat independent, but it is influenced by solar and temperature levels.

Of all factors that influence drying, solar radiation has the greatest impact. Drying rate has a higher and more consistent correlation with the level of solar radiation than any other crop or environmental variable (Table 3–1). Field curing of hay is very slow and perhaps impossible without the sun. Long-term drying data from humid climates indicate a 10-fold increase in drying rate between minimum and maximum solar radiation levels when all other variables are held constant (Table 3–2). Beside the direct effect of low solar radiation, lack of sunshine also leads to lower air temperatures and higher relative humidities that further reduce drying.

Solar radiation provides a vast source of inexpensive energy for drying the forage crop. Freshly mowed forage absorbs ≈80% of the incident radiation. As the crop dries, the solar absorptivity decreases, but it remains near 50% for the major portion of the drying time (Ajibola et al., 1980). Most of the solar energy is absorbed near the upper surface of the swath. In a relatively thick swath, radiation 2 cm below the surface is one-half that at the surface, and at the base it is ≈10% of the surface value (Jones & Harris, 1979). The absorbed energy is primarily used to evaporate moisture and secondarily to heat the crop.

The effect of solar radiation is interrelated with the structure of the drying swath. When the forage crop is dried in a thick swath, a smaller portion of the crop is exposed to the sun's radiant energy. With less exposure, the level of solar radiation has less direct effect on drying.

Solar radiation is sometimes viewed as less important when drying hay in dry climates. This is not because solar radiation is less necessary, but because it is more consistent from day to day in these climates. With little difference in solar radiation levels, differences in temperature and humidity have a more no-

Table 3-1. Partial correlations of field drying rate with various crop and environmental variables for data collected over several years of alfalfa production in Michigan (Rotz & Chen, 1985).

Environmental or crop variable	Correlation coefficient by data set[†]		
	Minimum	Maximum	Average
Solar insolation	0.51	0.70	0.61
Swath surface temperature	0.35	0.54	0.45
Ambient air temperature	0.30	0.45	0.35
Vapor pressure deficit	0.25	0.48	0.34
Crop moisture content	0.04	0.48	0.22
Relative humidity	−0.04	−0.38	−0.21
Swath density	−0.15	−0.20	−0.18
First day of drying indicator	0.06	0.30	0.19
Vapor pressure	0.08	0.28	0.15
Soil moisture content	−0.06	−0.26	−0.15
Wind velocity	0.07	0.17	0.11
Alfalfa cutting number	0.00	0.13	0.06
Alfalfa maturity	−0.11	0.06	0.00

[†] Correlation coefficients measured in five data sets collected between 1978 and 1984.

ticeable impact on drying. In dry climates, forage is often cured in relatively thick swaths or windrows to reduce bleaching of the crop by the sun. Solar radiation is not used as effectively and the crop dries more slowly.

The next most important weather variable affecting drying is vapor pressure deficit. This gradient in vapor pressure is controlled by plant temperature and the relative humidity of the surrounding air. Without receiving radiant energy, the plant temperature remains near that of ambient air. Incident solar radiation increases this temperature up to 20°C above that of ambient air (Dernedde, 1979). The vapor pressure deficit controlling drying, therefore, is related to am-

Table 3-2. Long-term average influence of environmental and crop variables upon alfalfa drying rate and field curing time in Michigan (Rotz & Chen, 1985).

Environmental or crop variable	Range in value		Drying rate[†]		Maximum effect on field curing time[‡]
	Minimum	Maximum	Minimum	Maximum	
			h^{-1}		h
Solar irradiance, W m^{-2}	100	950	0.046	0.232	48.0
Vapor pressure deficit, kPa	0	4.5	0.154	0.197	3.9
Air temperature, °C	10	40	0.153	0.186	3.2
Swath density, g m^{-2}	150	1500	0.199	0.128	7.8
Soil moisture content, % dry basis	10	25	0.196	0.160	3.2
First day of curing	0	1	0.156	0.178	2.3

[†] Drying rate expressed as portion of available moisture lost each hour.
[‡] Difference in field curing time (drying hours with rewetting and night periods excluded) between the minimum and maximum expected value of the crop or environmental variable.

bient air temperature, but it is often influenced by solar radiation. Long-term studies of field cured alfalfa indicate that with all other factors held constant, drying rate fluctuates 28% for a change in vapor pressure deficit of 4.5 kPa or 22% for a 30°C change in ambient air temperature (Table 3–2).

Relative humidity has a small effect on drying under good drying conditions. On sunny, warm days with relative humidities below 60%, the vapor pressure deficit is high enough to allow adequate drying. Lower humidities may allow better drying; however, plant and swath restraints often cause a greater constraint on drying than that imposed by humidity. As a result, field drying studies have shown little relationship between drying rate and the relative humidity of ambient air (Table 3–1; Rotz & Chen, 1985; Carlson et al., 1989).

When drying conditions are poor (low solar radiation and temperature) relative humidity can have a greater constraint on drying. With low radiant energy, the crop and the surrounding air temperatures are similar. With little temperature difference and a high relative humidity in the ambient air, a low vapor pressure deficit becomes the major constraint to drying. Field drying can be very slow on heavily overcast, humid days.

Wind can influence drying, but under most conditions the effect is small. Forage dried in a relatively thin swath laying close to the soil surface shows no correlation between drying rate and wind velocity (Table 3–1; Rotz & Chen, 1985). Wind velocity decreases toward the field surface with little air movement near the surface. Freshly mowed forage laid in a swath tends to collapse into a relatively dense swath <15 cm from the soil surface. Even on windy days, mean airflow through the swath rarely exceeds 0.2 m s^{-1} (Jones & Harris, 1979). The boundary layer around the plant and the static humid air trapped in the swath are influenced little by the wind.

More benefit can be obtained from wind by raking the forage off the field surface into a fluffy windrow. Air can then pass through the windrow carrying the moisture into the surrounding environment. The humidity inside the windrow can approach that of the ambient air, particularly when the crop is nearly dry (Jones & Harris, 1979). Following raking, the forage material settles back into a more dense swath, again restricting drying. When the crop is relatively dry at the time of raking, less settling occurs and more benefit is obtained for a longer time.

Latent and pan evaporation are other meteorological measures of drying conditions. Latent evaporation, measured with an atmometer, is the rate of evaporation from a wet, horizontal black porous surface. Pan evaporation measures the moisture that evaporates from free surface water on a given day. These measures integrate the effects of solar radiation, temperature, humidity, and wind. Forage drying is related to these measures of evaporation (Kemp et al., 1972; Pitt, 1984), but crop constraints must be considered. Early in the drying process, moisture loss from the crop is close to that from free surface water. As the crop dries, the resistances imposed by the plant tissue, the surrounding boundary layer, and the moist air enclosed in the swath restrain drying, so a measure of moisture evaporation has less relationship to drying.

Soil is another part of the environment that affects drying. Soil moisture is the predominant factor, but soil temperature can have very minor effects. Soil moisture and temperature control the vapor pressure under the swath. On wet soil, the vapor pressure near the soil can be greater than that in the swath causing moisture migration from the soil to the swath. Humid air within the swath restricts drying and under severe cases may contribute to a gain in forage moisture. A change from the minimum to the maximum soil moisture expected under normal conditions causes about a 20% decrease in the drying rate of field cured alfalfa (Table 3–2).

Soil moisture effects on drying can be controlled. Placing a moisture barrier between the swath and soil can greatly increase the rate and extent of drying (Pedersen & Buchele, 1960a). Raking the swath into a narrow, more fluffy windrow can improve drying by reducing the amount of crop directly exposed to wet soil. Laying the swath on a long stubble also may reduce the effect of soil moisture. A uniform stubble of at least 50 to 75 mm in length is important for the ventilation of swaths and as a barrier to soil-moisture uptake (Klinner & Shepperson, 1975).

Equilibrium Moisture Content

Equilibrium moisture content is the moisture content the forage plant attains if placed in a constant environment for an indefinite period of time. This moisture level is primarily related to the environment and secondarily related to the plant (Bakker-Arkema et al., 1962). Equilibrium moisture is influenced most by the relative humidity of the surrounding air with the greatest influence at high humidities. Within the range of relative humidities existing during good drying days (20 to 60%), equilibrium moisture content increases only slightly with an increase in humidity. For relative humidities above 70%, equilibrium moisture increases rapidly (Fig. 3–4).

Air temperature also affects equilibrium moisture. For temperatures experienced during field curing, the effect is small unless the relative humidity is <60% (Fig. 3–4). At temperatures >90°C, equilibrium moisture content is essentially zero (Hammer & Day, 1967).

In a given environment, the equilibrium moisture content is somewhat greater if the crop is losing moisture than when it is absorbing moisture. This difference, known as the hysteresis effect, is more pronounced at high relative humidities (Bakker-Arkema et al., 1962). For alfalfa dried in humidities <70%, the difference is <30 g kg^{-1} (wet basis, wb), but at higher humidities, the difference is as much as 100 g kg^{-1} (wb).

Crop maturity also may affect equilibrium moisture content. Equilibrium moisture of pre-bloom alfalfa is ≈40 g kg^{-1} (wb) greater than that for full-bloom alfalfa (Bakker-Arkema et al., 1962). For very high humidities this difference is >100 g kg^{-1}.

Radiant energy appears to reduce the effective equilibrium moisture content of forage. When dried in the field exposed to solar radiation, the crop dries rapidly as though drying toward a moisture content of 0 g kg^{-1} (Rotz & Chen, 1985; Ajibola et al., 1980). This does not necessarily mean that forage left in the field will dry to a point where there is no moisture remaining in the plant. The equilibrium moisture is low, and it has less relationship with the ambient temperature and humidity. This is at least partly due to an increase in plant tempera-

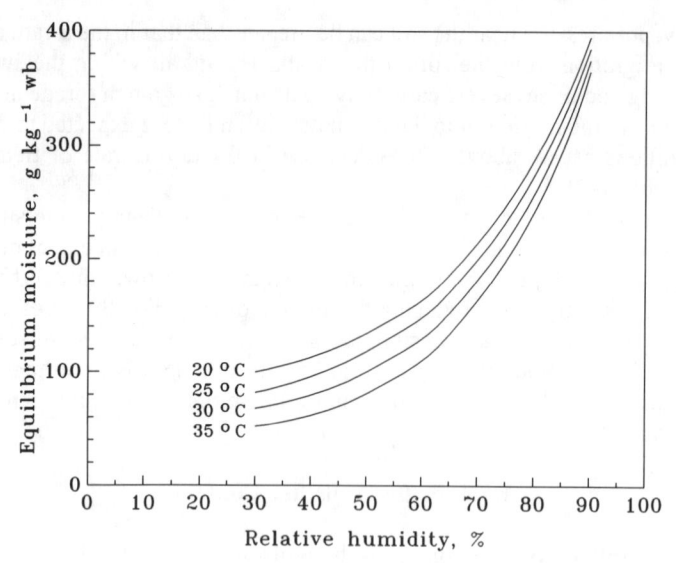

Fig. 3–4. Equilibrium moisture content of alfalfa expressed on a wet basis (wb; Hill et al., 1977).

ture produced by the absorbed energy. The elevated temperature and additional energy absorbed drive more moisture from the plant than would otherwise occur.

FACTORS INFLUENCING THE REWETTING OF FORAGE

Rewetting occurs during the field curing of most forage crops. In dry climates, rewetting is often used to advantage to reduce baling losses. Hay is baled at night after very dry hay has absorbed moisture from the humid air. With the added moisture, leaves are less brittle and less likely to be damaged or lost during baling. In all other cases, rewetting is undesirable, prolonging the field curing process and promoting deterioration of the crop.

Forage rewetting occurs by: (i) absorption of moisture from humid air, (ii) dew formation, and (iii) rainfall. There are similarities between these three processes; however, the crop and environmental factors that affect the rate and extent of rewetting are quite different.

Moisture Absorption From Humid Air

As described above, moisture moves between the mown plant and its environment drawing them toward equilibrium. When the vapor pressure in the environment is greater than that in the plant, moisture is absorbed by the plant. Moisture absorption most often occurs at night, particularly when the crop is relatively dry. Moisture absorption also may occur with a dry crop on an overcast, humid day.

Moisture absorption occurs because the equilibrium moisture content for the crop is greater than the actual moisture content (Fig. 3–4). In order for this to occur, the relative humidity of the ambient air must be very high (80 to 100%) or the crop must be very dry. The rate of moisture absorption is controlled by the same crop and environmental factors that control drying. The boundary layer of air surrounding the plant, the plant cuticle and the plant structure restrict the migration of moisture back into the plant. Although the field drying process is well documented, much less is known about absorption during field curing.

Crop moisture content has the greatest effect on moisture absorption. The moisture absorbed per unit of crop dry matter increases as the moisture content of the crop entering the rewetting cycle decreases. When the crop is relatively dry, the gradient between the crop moisture and the equilibrium moisture content is greatest, which causes more rapid rewetting and more moisture absorption.

Swath structure affects moisture absorption. A thin swath spread over much of the field surface is more fully exposed to the humid conditions. Likewise, a more open, fluffy swath or windrow may absorb moisture a little more quickly than a thick, dense swath. With less resistance between the plant and its environment, moisture is absorbed more rapidly.

The physical characteristics of the plant that influence drying also may influence the rate of absorption. Theoretically, forage with a high leaf to stem ratio will absorb moisture more rapidly due to the greater surface area per unit volume. Likewise, forage with thin stems may absorb moisture more quickly than thicker stems. Treatments applied to the plant to improve drying may enhance rewetting, because breaks in the plant cuticle can allow moisture to move back into the plant more easily. The effects of these physical characteristics on the rewetting of field cured forage have not been demonstrated. For most conditions, they appear small and relatively unimportant in light of other factors.

Environmental factors that affect moisture absorption include relative humidity, temperature and wind. Equilibrium moisture content is ≈ 200 g kg^{-1} (wb) when the relatively humidity of the ambient air is 70%, but it increases rapidly to >400 g kg^{-1} for humidities >90% (Fig. 3–4). Temperature of the ambient air has less effect with only a small increase in equilibrium moisture as temperature decreases. A high equilibrium moisture increases the vapor pressure gradient and thus increases the rate and extent of absorption. Wind normally has little effect on moisture absorption.

Dew Formation

Moisture absorption at night is often augmented by dew formation. Moisture forming on the plant surface is absorbed into the dry internal tissue. The rate of moisture migration into the plant is again related to plant characteristics such as cuticle thickness, leaf to stem ratio and stem thickness. The maximum amount of dew that can be absorbed by forage plants is not documented, but it is likely close to the initial moisture content of the crop. In addition, surface moisture remaining from dew must be removed. Surface moisture evaporates relatively fast when drying resumes, but moisture absorbed into the plant dries at a rate similar or slightly greater than that of the original plant moisture (Pitt & McGechan, 1987; Jones & Palmer, 1932).

The duration and extent of dew formation is related to air temperature, humidity, wind speed, and cloud cover (Pedro & Gillespie, 1982). Dew forms when the temperature of the plant drops below the dew point of the ambient air.

When saturated air cools, less vapor can be contained so a portion condenses to a liquid state. Therefore, when air saturated with vapor comes in contact with forage plants at a lower temperature, the air cools and condensation occurs. Thus for dew to occur on field curing forage, the plant temperature must drop below the ambient air temperature. Long wave radiation to the earth's outer atmosphere removes heat from the plant material, dropping the temperature below that of the surrounding air. The amount of energy radiated to the atmosphere is influenced by cloud cover. A clear sky absorbs more radiant energy giving lower plant temperatures and a greater rate and extent of dew formation. The duration of dew is directly related to the interval of clear night sky conducive to dew formation. The amount of dew formed on the plant is directly related to the length of the dew duration, but the rate of absorption decreases as plant moisture content increases.

The relative humidity of ambient air also must be high for dew to form. As humidity increases, the dew point approaches the ambient temperature, so less temperature drop is required to cause dew. Among the factors that control dew, the nighttime duration of relative humidity >90% exhibits the strongest correlation with dew formation (Crowe et al., 1978).

Wind also has a relatively large effect on dew. In order for the plant temperature to drop below the ambient air temperature, the air around the plant must be relatively still. When the surrounding air is moving, energy is absorbed from the air, and the plant temperature remains close to that of the ambient air. Therefore, wind can reduce and even eliminate dew formation.

Swath structure also can influence rewetting from dew. A thin wide swath exposes more crop to long wave radiation, and the dew that forms. In a thick, narrow swath, only the upper surface is exposed so there is less rewetting of the total crop. Proper timing of raking operations is important to maximize drying and minimize rewetting for optimum field curing.

Rainfall

The most extensive rewetting of forage is caused by rain during the field curing process. Rain adheres to the surface of plant material where a portion of the moisture is absorbed into the inner plant tissue. When rain occurs, a portion of the moisture moves through the swath to the soil below. The moisture retained can be categorized as: (i) moisture absorbed into the plant, (ii) moisture adhered to the plant surface, and (iii) moisture temporarily held by the plant before running off (van Elderen et al., 1972). The amount of moisture retained is a function of the amount, duration, and intensity of the rainfall and the crop and swath characteristics. The maximum amount retained is 9 kg of moisture per kg of crop dry matter (van Elderen et al., 1972).

Of the moisture retained by the crop, that absorbed into the plant tissue slows field curing the most. Excess moisture runs off the crop soon after rain ceases. The free water retained on the surface evaporates relatively easily when drying resumes, unless that moisture is embedded in a thick swath. Absorbed moisture dries more slowly though, because the moisture must migrate back to the plant surface. Given the same drying conditions, absorbed rain moisture may dry a little faster than the original tissue moisture (Pitt & McGechan, 1987).

The amount of moisture absorbed is primarily influenced by the amount of rainfall and secondarily by the intensity and duration of rainfall. With a small amount of rain, nearly all moisture falling on the crop is retained. When the specific rainfall (kilogram of incident rain per kilogram of crop dry matter) exceeds 2.0, runoff begins (Pitt & McGechan, 1987). Moisture retention also is influenced by rain intensity. Less moisture is retained as the intensity of the rain increases (Fig. 3–5). Slow rainfall over longer durations allows more moisture to be absorbed into the plant material. Therefore, moisture retention early in rainfall is nearly 100%, but as rain continues the portion retained drops to zero as the maximum possible moisture content for the crop is met.

Rain retention also is influenced by the swath structure. With a light rain, a wide swath retains more moisture than a narrow swath. The wide swath exposes more crop area to rain, i.e., the specific rainfall is increased. As the amount and duration of rainfall increase, swath structure has less influence. Although a wide swath may rewet faster, a similar extent of rewetting occurs regardless of swath structure when excessive amounts of moisture are added.

Plant characteristics also may influence the rate and extent of rewetting, but such influences have not been well documented. A crop with a high leaf-to-stem ratio likely retains more moisture per unit of crop dry matter due to the greater surface area. Leaves also may absorb moisture more quickly than stems. Treatments applied to the plants to speed drying also may affect moisture absorption. In one study, mechanically conditioned forage retained less moisture than nonconditioned forage (Pitt & McGechen, 1987).

METHODS FOR SPEEDING FIELD CURING

Perhaps as long as hay has been made, producers have wanted to speed the field curing process. Reducing field curing time improves the timeliness of forage harvest and lessens the dry matter and nutrient losses during harvest. Many methods have been used to speed hay drying. The oldest is some method of turning or manipulating the swath during drying. In the past 50 yr, mechanical conditioning has become a common practice, and more recently the concept of chemical conditioning was introduced. Other methods have investigated the use of heat to disrupt the plant cuticle. Exposure of forage plants to an open flame (Person & Sorenson, 1970), steam (Byers & Routley, 1966), heated rolls, a hot water blanch, microwave radiation (Priepke & Bruhn, 1970), or high voltage electrical currents (Sirohi et al., 1985) provides small to moderate improvements in drying. Field application of these treatments is not practical due to the need for unreasonable handling of materials, excessive energy, and expensive equipment.

Swath Manipulation

As forage cures in the field, the top of the swath dries more rapidly than the bottom. Manipulation of the swath, therefore, can speed the drying process by

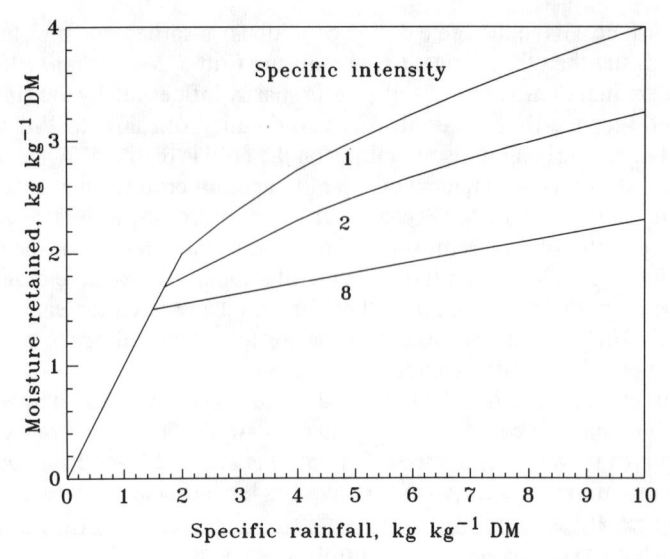

Fig. 3–5. Moisture retained vs. specific rainfall for three specific rainfall intensities expressed in kilograms of rain per kilogram of dry matter (DM) per hour (Pitt & McGechan, 1987).

moving the wetter material to the upper surface. Swath manipulation also can improve drying by spreading the hay over more of the field surface. Spreading exposes more of the crop to the incident solar radiation and ambient air to enhance evaporation and diffusion of moisture.

Tedding, swath inversion and raking are used in haymaking to manipulate the swath. A tedder uses rotating tines to stir, spread, and fluff the swath. Swath inverters gently lift and turn the swath placing it back on the field surface inverted from its original position. For raking, tines are used to sweep or roll a swath into a more narrow windrow or to turn a windrow.

The drying rate increase obtained with tedding is related to the way the tedder is used. The stirring or fluffing aspect of tedding provides a 20 to 40% increase in drying rate on the day the treatment is applied with an average increase of 30% (Pattey et al., 1988; Savoie & Beauregard, 1990a). This increase can reduce the field curing time by 2 to 8 h. Tedding is often used to spread a narrow swath formed by the mower-conditioner over the entire field surface. When spread within a few hours after mowing, the drying rate is increased 30 to 40% for the entire field curing period, and that reduces field curing time by ≈1 d. Similar drying results can be obtained with both legume and grass crops. Tedding is most effective on high-moisture, early-maturity hay. In addition to speeding the drying process, tedding also may allow more uniform drying.

Disadvantages of tedding include an increase in fuel, labor, and machinery costs and an increase in leaf shatter. Tedding wet alfalfa causes a small yield loss of ≈1%. Greater loss occurs in a drier crop, with as much as 10% loss in alfalfa at 300 g kg^{-1} moisture. In grass crops, tedding losses are much less, about one-quarter that of alfalfa (Savoie, 1987). After a crop is spread with a tedder, it is more difficult to gather with a rake, which leads to additional raking loss. The decision to use tedding must be made by comparing the probable loss from rain damage to the known loss and cost of tedding (Rotz & Savoie, 1991). If tedding is used, the best time to ted the crop to reduce loss and obtain the best drying benefit is early in the drying process, following rainfall or early in the morning when the crop is wet with dew.

Swath inversion exposes the wet bottom layer of the swath for more rapid drying. Inversion increases the drying rate of alfalfa ≈15% on the day the treatment is applied, and this reduces field curing time by 1 to 4 h (Savoie & Beauregard, 1990b). The swath inverter also may spread the crop to provide some additional drying improvement. Overall, swath inversion is not as effective as tedding, but it has an advantage in that leaf shatter loss is low. The more gentle crop handling causes less than a 1% loss in alfalfa (Savoie & Beauregard, 1990a). Again, the added operation in haymaking increases labor, fuel, and machinery costs. The small improvement in drying generally does not justify the greater cost (Rotz & Savoie, 1991).

Raking, unlike the previous two operations, creates a narrower swath or windrow. Raking tends to roll the wetter hay from the bottom of the swath to the outer surface of the windrow. The hay is often rolled onto drier soil, which also improves drying. Raking provides a 10 to 20% increase in drying rate on the day the hay is raked. Following the initial improvement in drying, the thicker swath or windrow can restrain drying. Therefore, the moisture content of the crop at the time of raking is important. If the crop is too wet, the wet material rolled into the center of the windrow dries slowly.

Raking also causes loss, and the loss is related to crop moisture. Typical losses range from 2% in moist crops to 15% in very dry legume crops. The best time to rake for minimum loss and optimum drying is when the crop is between 300 and 400 g kg^{-1} moisture. Raking in the morning of the day in which hay is anticipated to be ready for baling can reduce field curing time by a few hours, thus allowing an earlier start at baling.

Different rake designs are available and some are promoted to provide faster drying. Major types can be classified as parallel-bar, wheel, and rotary rakes. Parallel-bar and wheel rakes are similar in that they roll a swath into a more narrow windrow. As the forage is rolled, the wetter material from the bottom of the swath is wrapped on the outside of the windrow. A rotary rake uses horizontal rotating tines to sweep the swath into a windrow. The sweeping action tends to create a less dense windrow, but losses are greater. When swept across the stubble, more forage particles are lodged in the stubble. Substantial and consistent differences in drying have not been measured between the major rake designs (Savoie et al., 1982).

In humid climates, hay can be dried in relatively wide, thin swaths and raked in the morning, several hours before baling. This procedure provides relatively fast drying with minimal shatter loss. Additional manipulation of the swath can speed drying particularly following rainfall. The cost for routine use of tedders and swath inverters is difficult to justify on legume crops (Rotz & Savoie, 1991). For grass crops grown in very humid climates, tedding may be beneficial in hay making, but not in silage making (McGechan, 1990). When harvesting silage or low-yielding hay, the best general procedure is to adjust the mower-conditioner for a narrow swath and then allow the swath to dry without further manipulation until the crop is chopped or baled.

Mechanical Conditioning

A mechanical treatment used to reduce the forage plant's resistance to moisture loss is commonly known as conditioning. Conditioning devices speed drying by crimping, crushing, abrading, or macerating the plant. A crimping device passes the crop between intermeshing, noncontacting rolls that break the stem at ≈5-cm intervals. Crushing smashes the plant between smooth rolls causing longitudinal splitting of the stem. Most current roll conditioning devices use intermeshing rolls that both crush and crimp the crop. Plant moisture evaporates more easily from these splits and breaks in the epidermis. Other conditioning equipment use abrasion to remove, scratch, or otherwise disturb the cuticle of the forage plant to allow moisture to pass more easily. Maceration is the ultimate form of conditioning. With maceration the crop is shredded, fully exposing internal plant tissue for rapid drying.

Although many forms of conditioning consistently provide some improvement in drying, the amount of improvement obtained varies widely. A high degree of interaction occurs between the effectiveness of a conditioning treatment and the plant, swath, and environmental conditions. Crop conditioning is most effective when the plant is the primary limit to drying. For this to occur, a thin or fluffy swath is needed and the weather conditions must be good to promote fast drying. Mechanical conditioning of the plant has little or no benefit when other drying conditions are poor.

Differences among forage drying treatments are best obtained in an ideal drying environment. Laboratory drying studies often compare treatments by moving air through a sample of forage. Greater differences among conditioning treatments are obtained than can be acquired in the field where the swath and environment play more important roles in restricting drying.

Conditioning treatments also are more effective on some plant material than others. Forage with a large stem and/or a low leaf-to-stem ratio responds best to conditioning treatments. For this reason, mechanical conditioning of alfalfa in Michigan increased drying rate during first cutting by 80% with a small effect on second cutting and no effect on later cuttings (Table 3–3). Mechanical treatments also tend to be more effective on legumes in an early stage of maturity (Bruhn, 1955). As the crop matures, a decrease in the initial moisture content, a hallowing of stems, and perhaps a breakdown or opening of the cuticle improves drying, so conditioning has less opportunity to further improve drying.

Conditioning may cause greater rewetting of the crop during periods of high humidity, dew, and light rain (Kepner et al., 1960; Fairbanks & Thierstein, 1966). This occurs partly because a conditioned crop is drier when these conditions begin. Under a given set of conditions, a drier crop absorbs more moisture. The disrupted plant cuticle also may allow moisture to move back in the plant more easily.

Conditioning devices cannot be selected solely on the basis of drying improvement. Dry matter losses and the associated nutrient changes caused or pro-

Table 3-3. Typical drying rates and field curing times for alfalfa dried in a wide swath following various conditioning treatments in Michigan (Rotz et al., 1987).

Cutting	Conditioning treatment	Drying rate† Average	Drying rate† Increase	Field curing time‡
		h^{-1}	%	daytime h
1	None	0.07	--	40
	Mechanical	0.12	79	23
	Chemical	0.10	43	28
	Mechanical & chemical	0.15	114	19
2	None	0.11	--	24
	Mechanical	0.15	36	18
	Chemical	0.23	120	11
	Mechanical & chemical	0.24	130	10
3	None	0.12	--	20
	Mechanical	0.12	0	20
	Chemical	0.19	57	13
	Mechanical & chemical	0.19	57	13
4§	None	0.08	--	35
	Mechanical	0.08	0	35
	Chemical	0.09	20	28
	Mechanical & chemical	0.09	20	28

† Drying rate expressed as portion of available moisture lost each hour.
‡ Drying time to 200 g kg^{-1} moisture content (wet basis) with rewetting and night periods excluded.
§ Fourth cutting taken about 15 October.

moted by the treatment also must be considered. In general, losses increase with the severity of conditioning. Although heavy conditioning can provide the fastest field curing, the forage is more susceptible to high harvest losses.

Many different roll designs are used in today's mower-conditioners. Some designs are promoted as providing faster drying, but this is not supported by research. In general, crushing treatments are more effective than crimping treatments. Some of the earliest conditioner designs used rolls that only crimped the crop, but this type of roll is not used today. Field and laboratory drying studies consistently show little or no difference in the drying of legumes or grasses treated with various crushing roll designs (Bruhn, 1955; Straub & Bruhn, 1975; Rotz & Sprott, 1984; Shinners et al., 1991).

Factors such as roll pressure, roll clearance, and feed rate have more effect on drying than the type of roll. Roll pressure is the amount of pressure placed on the crop as it passes between the rolls. More pressure causes more rupturing of plant tissue and thus faster drying (Bruhn, 1955). Roll clearance also can affect the pressure imposed on the crop. A machine with a small, fixed clearance provides severe crushing and rapid drying (Pedersen & Buchele, 1960b). This type of design is not used because it also causes greater loss.

The feed rate of crop between the rolls affects the amount of conditioning received. At a low feed rate, a thin mat passing between the rolls tends to receive more crushing damage. Feed rate is controlled by the peripheral roll speed and the length of the rolls relative to the machine's mowing width. When the rolls are rotated faster than the ground speed of the machine, the crop is drawn into a thinner mat passing through the rolls. Likewise when the rolls run the full width of the machine, a thinner mat is created. The effect of feed rate is small and difficult to detect in field drying studies.

Flail mowing machines have been used for many years to mow and condition forage crops. Rotating flail blades mow the crop and provide additional cutting and abrasion as the crop moves around the flail mechanism back to the field surface. Proper adjustment of flail speed relative to ground speed is critical. When conditions are right, flail mower conditioners provide faster drying than can be obtained from roll machines (Rotz & Sprott, 1984). With less ideal situations, the flail causes excessive conditioning with shorter pieces of limp crop that settle into the crop stubble in a dense swath (Boyd, 1959). Other disadvantages of flail mower-conditioners are greater field losses and greater power requirements compared with roll type machines (Rotz & Sprott, 1984).

Machines that condition the crop with abrasion can be more effective on grass crops than roll-type machines (Klinner, 1976). These machines use rotating brushes or tines to scratch or abrade the plant cuticle. Since the stems are not broken or ruptured, they remain more stiff creating a less dense swath or windrow. This type of conditioner provides a greater throughput capacity when harvesting very high yielding or entangled crops that are difficult to handle with roll type mower-conditioners. Conditioning of alfalfa with rotating plastic tines can provide a drying rate substantially less than that obtained with roll conditioning (Rotz & Sprott, 1984).

Maceration is the most severe form of mechanical conditioning. Plant stems are shredded, fully exposing the internal moisture. Drying restraints imposed by the internal cell structure, the epidermis and the cuticle are removed allowing moisture to evaporate similar to that from free surface water. Macerated alfalfa can dry to a moisture content suitable for baling in 4 to 6 h of favorable drying conditions (Krutz & Holt, 1979).

Macerated forage contains many fine particles that are very susceptible to loss during field curing and harvest. To reduce the potential loss, the shredded material can be pressed into a mat that is laid on the field surface for rapid curing (Koegel et al., 1988). The dried mat can then be picked up with minimal loss. Commercial equipment for maceration and mat drying of alfalfa is not yet available, but research and development of the process continues. In order to implement the process on the farm, new equipment and procedures are required for forage harvesting, handling, and storage.

Fast drying is a major advantage of the mat process. The drying rate of alfalfa mats is two to three times greater than the rate of conventional swaths. A major factor affecting mat drying (other than the environment) is the mat thickness. A relatively thin mat (<1.5 cm) must be maintained because moisture on the bottom of a thicker mat dries very slowly. With an appropriately designed mat system, hay can be made with 1 d of field curing. Disadvantages of the mat system include high power requirements and equipment costs. The long-term benefits of faster drying, reduced loss, and improved forage quality appear to outweigh the added fuel and equipment costs (Rotz et al., 1990).

Chemical Drying Aids

Chemicals can be used to speed the field curing of both legume and grass forage crops. Chemical treatments improve drying by opening stomata, desiccating the plant prior to mowing, or by modifying the epicuticular waxes. Stomata can be held open or opened by treatments of sodium azide, fusicoccin, and kinetin (Harris & Tulberg, 1980). In laboratory studies, these materials improved the drying of leaves from various forage species, but practical systems for field use were not considered. Plant desiccation can be done by applying certain herbicides prior to mowing. Dinoseb [2-(sec-butyl)-4,6-dinitrophenol], endothal (7-oxabicyclo[2,2,1]heptane-2,3-dicarboxylic acid), and diquat (1,1'-ethylene-2,2'bipyridylium ion) treatments improved drying primarily by reducing the moisture content of the standing crop by 60 to 70 g kg^{-1} (Harris & Tulberg, 1980). Plant discoloration, chemical costs, and concern for chemical residues have prevented commercial use of herbicides to speed drying.

Other chemical treatments modify the epicuticular waxes to improve forage drying. In the laboratory, exposure to petroleum ether or benzene caused marked changes in the surface waxes with large increases in forage drying rates (Jones & Harris, 1979; Morris, 1972). Application of propionic and formic acids provided small and inconsistent improvements in field curing of grass crops (Harris & Tulberg, 1980).

The first practical method for using chemicals to speed the field curing of forage came with the use of aqueous potassium carbonate. The exact mechanism by which this treatment improves drying is unknown. The solution is thought to form a continuous film over the plant surface that extends down the cavities between the wax platelets. The film joins the liquid phase moisture within the parenchyma, pectin, and cuticular membrane, and thus enables moisture transfer in the liquid phase through the wax layer by capillary forces (Harris & Tulberg, 1980). When sprayed on the crop as it is mowed, the solution can increase drying rate. This process of chemical conditioning is effective on legume forage species, but not on grass species (Johnson et al., 1984).

Potassium carbonate is the most common active ingredient in chemical drying agents. A 28 g L^{-1} solution of K$_2$CO$_3$ in water applied at an appropriate rate speeds drying. A mixture of K$_2$CO$_3$ and Na$_2$CO$_3$ is less expensive, but equally effective to K$_2$CO$_3$ alone (Rotz & Thomas, 1988). Most commercially available drying agents are approximately a 50:50 mixture of K$_2$CO$_3$ and Na$_2$CO$_3$. The treatment is more effective as more solution is applied, but the economically optimum amount is ≈300 L ha^{-1} for yields under 3.5 t ha^{-1} and 470 L ha^{-1} for greater yields. Mixing more chemical per unit of water does not improve drying performance nor compensate for a decrease in solution application rate (Rotz & Davis, 1986).

The best method for chemical application is through a spray system mounted on the mower conditioner. Normally a spray boom is mounted just ahead of the cutterbar to spray the crop as it is cut. A pushbar is used ahead of the spray boom to open up the plant canopy to allow the spray to cover the entire plant. On some mower-conditioners, the spray boom can be mounted just ahead of the conditioning rolls to spray the mat of mown alfalfa as it enters the rolls. When a roll-type conditioner is used, the rolls help spread the chemical over the plant, which may improve the treatment effectiveness. The type of nozzle used is not important as long as the correct application rate is obtained with relatively uniform coverage across the width of the machine (Rotz & Davis, 1986). Use of fine droplet atomizers does not improve the effectiveness of the treatment nor reduce the required application rate of solution (Rotz & Davis, 1987).

The effectiveness of a chemical drying agent increases as drying conditions improve. Chemical treatment provides the greatest increase in drying rate when the forage plant is the predominant restraint to drying. This occurs when the crop is dried in thin, loose swaths on sunny, warm days. Chemical conditioning is less effective on first cutting alfalfa because greater crop yields create thicker swaths. In thick swaths, the swath becomes the major restraint to drying, and the resistance of the plant to moisture removal has less effect on overall drying time.

Chemical conditioning can improve the drying of all cuttings of alfalfa, but it is most effective on summer harvests (Rotz et al., 1987). In the Midwest, chemical conditioning provided about a 40% increase in the drying rate of first cutting alfalfa with up to a 120% increase on second and third cuttings (Table 3–3). Poor drying conditions in the fall limit the increase to ≈20%.

The effects of mechanical and chemical conditioning are independent, i.e., chemical conditioning affects the crop in a different way than mechanical conditioning. When chemical conditioning is used, a similar increase in drying occurs whether it is used with or without mechanical conditioning (Table 3–3). Although the absolute increase in drying rate is similar, the relative increase is less when chemical conditioning is added to mechanical conditioning. For example, an increase in the drying rate of 0.03 h^{-1} obtained with chemical conditioning is a 40% increase in drying rate over no conditioning, but only a 25% increase over mechanical conditioning (Table 3–3, Cutting 1). In the latter case, the 25% increase in drying rate provides an additional reduction of field drying time of 4 h over that obtained with mechanical conditioning. On later cuttings where mechanical conditioning is less effective, the combination of both is the same as chemical conditioning alone. The typical reduction in field curing time for chemical conditioning compared with mechanical conditioning is 5 daytime h on first cutting and 7 to 8 h on later cuttings (Table 3–3).

SUMMARY

Field curing of forage involves diurnal cycles of drying and rewetting until a moisture content suitable for harvest is attained. Energy from the sun or ambient air evaporates moisture from forage plants as they lay on the field surface. The moisture diffuses in the atmosphere moving toward a lower vapor pressure in the surrounding air. As conditions change, moisture from rain, dew, or very humid air is absorbed causing a rewetting of the crop.

The drying rate of forage crops is restricted by plant, swath, and environmental factors. The restraint each places on drying varies in relation to the others. Most often, the environment is the primary restraint. High levels of solar radiation, high temperatures, low humidities, and low soil moistures all promote good drying. When environmental conditions are good, a thick, dense swath may inhibit drying. Under this scenario, drying is enhanced by treatments that turn, fluff, or spread the swath. Mechanical and chemical treatments that alter the plant can aid drying when environmental and swath conditions are favorable, but they cannot compensate for poor drying conditions.

Swath manipulation with a tedder, inverter, or rake improves drying, but the improvement is not normally large. Tedding and inverting are not justified for routine use on legume crops; however, they may be beneficial following rain or extended poor drying conditions. For any of these operations, timing is critical to minimize loss and maximize the drying benefit.

Mechanical conditioning can speed the field curing of all forage crops. Rolls that crush and crimp the crop provide the best conditioning of legume forages. The crushing and crimping treatment is most effective on crops with a thick stem and low leaf-to-stem ratio such as occurs in first cutting alfalfa. Most roll types provide similar improvements in drying. For alfalfa grown in a four cutting system in humid climates, mechanical conditioning can increase the drying rate of first cutting by 80% with a smaller effect on second cutting and little effect on third and fourth cuttings. For grass forage crops, devices that use rotating tines, flails, or brushes to abrade the plant surface provide good drying improvements. The abrasion disrupts the plant cuticle yet maintains the strength of the stem to give a less dense swath or windrow.

Chemical treatments can be used to enhance drying by opening stomata, disrupting the plant cuticle, or improving moisture migration through the cuticle. The latter method is used to improve the drying of legume crops by applying a 28 g L^{-1} aqueous solution of K_2CO_3 or similar chemical at mowing. The best drying improvement is again obtained under good weather and swath drying conditions. When properly used, chemical conditioning can reduce field curing time by up to 1 d. Grass drying rates also can be improved by chemical treatments, but practical treatments for farm use have not been widely developed.

Maceration or shredding provides the ultimate form of forage conditioning. Shredding fully exposes internal plant tissue for rapid drying, but the shredded crop is very susceptible to loss. When pressed into a mat, the forage can be laid back on the field surface for rapid curing with little loss. Very high quality forage can be produced with 1 d of field curing. More research and development is required to create a process for commercial use.

Field curing time is only one of several considerations when evaluating and selecting forage harvest systems. The overall goal is to harvest forage with minimum nutrient loss at the lowest cost. Enhancing field curing is an important part of reducing field losses, but treatments that enhance drying often cause some loss as well. These treatments also may increase equipment, fuel, labor, and chemical costs. Only by balancing field curing benefits with the losses and costs incurred can the best system for forage harvest be selected. Best options for field curing vary with climatic regions, crops grown, and farm management styles. This information on field curing of forages is intended to aid the selection of harvest systems by providing a basic knowledge of the field curing process.

REFERENCES

Ajibola, O., R. Koegel, and H.D. Bruhn. 1980. Radiant energy and its relation to forage drying. Trans. ASAE 23:1297–1300.

Bagnall, L.O., W.F. Millier, and N.R. Scott. 1970. Drying the alfalfa stem. Trans. ASAE 13:232–236.

Bakker-Arkema, F.W., C.W. Hall, and E.J. Benne. 1962. Equilibrium moisture content of alfalfa. Michigan Agric. Exp. Stn. Quart. Bull. 44:492–496. East Lansing.

Boyd, M.M. 1959. Hay conditioning methods compared. Agric. Eng. 40:664–667.

Bruck, I.G.M., and E. van Elderen. 1969. Field drying of hay and wheat. J. Agric. Eng. Res. 14:105–116.

Bruhn, H.D. 1955. Status of hay crusher development. Agric. Eng. 36:165–170.

Byers, G.L., and D.G. Routley. 1966. Alfalfa drying, overcoming natural barriers. Agric. Eng. 47:476–485.

Carlson, J.D., H. Choi. and C.A. Rotz. 1989. The influence of weather variables on the drying of hay. p. 42–45. In Proc. 19th Conf. on Agric. and Forest Meteor., Charleston, SC. 7–10 Mar. 1989. Am. Meteorol. Soc., Boston, MA.

Collins, M., E. Baker, and N.L. Taylor. 1991. Trichome density effects on red clover hay drying. p. 65–67. In Proc. Am. Forage and Grassl. Conf., Am. Forage and Grassl. Counc., Columbia, MO. 1–4 Apr. 1991. Am. Forage and Grassl. Counc., Georgetown, TX.

Clark, B.A., and P. McDonald. 1977. The drying pattern of grass swaths in the field. J. Brit. Grassl. Soc. 32:77–81.

Clark, E.A., S.V. Crump, and S. Wijnheijmer. 1985. Morphological determinants of drying rate in forage legumes. p. 137–141. In Proc. Am. Forage and Grassl. Conf., Hershey, PA. 3–6 Mar. 1985. Am. Forage and Grassl. Counc., Georgetown, TX.

Crowe, M.J., S.M. Coakley, and R.G. Emge. 1978. Forecasting dew duration at Pendleton, Oregon, using simple weather observations. J. Appl. Meteorol. 17:1482–1487.

Dernedde, W. 1979. Treatments to increase the drying rate of cut forage. p. 61–66. In C. Thomas (ed.) Proc. Occasional Symp. 11, Forage Conservation in the 80s, Brighton, England. 27–30 Nov. 1979. Brit. Grassl. Soc., Grassl. Res. Inst., Hurley, England.

Fairbanks, G.E., and G.E. Thierstein. 1966. Performance of hay-conditioning machines. Trans. ASAE 9:182–184.

Galeano, R., M.D. Rumbaugh, D.A. Johnson, and J.L. Bushnell. 1986. Variation in epicuticular wax content of alfalfa cultivars and clones. Crop Sci. 26:703–706.

Hammer, P.C., and C.L. Day. 1967. Thin layer hay drying. Trans. ASAE 10:645–647.

Harris, C.E., and V.S. Shanmugalingam. 1982. The influence of the epidermis on the drying rate of red clover leaflets, leaf petioles and stems at low water contents. Grass Forage Sci. 37:151–157.

Harris, C.E., and J.N. Tulberg. 1980. Pathways of water loss from legumes and grasses cut for conservation. Grass Forage Sci. 35:1–11.

Hill, J.D., I.J. Ross, and B.J. Barfield. 1977. The use of vapor pressure deficit to predict drying time for alfalfa hay. Trans. ASAE 20:372–374.

Johnson, T.R., J.W. Thomas, C.A. Rotz, and M.B. Tesar. 1984. Drying of cut forages after spray treatments to hasten drying. J. Dairy Sci. 67:1745–1751.

Jones, L. 1979. The effect of stage of growth on the rate of drying of cut grass at 20°C. Grass Forage Sci. 34:139–144.

Jones, L., and C.E. Harris. 1979. Plant and swath limits to drying. p. 53–60. In C. Thomas (ed.) Proc. Occasional Symp. 11, Forage Conservation in the 80s, Brighton, England. 27–30 Nov. 1979. Brit. Grassl. Soc., Grassl. Res. Inst., Hurley, England.

Jones, T.N., and L.O. Palmer. 1932. Field curing of hay as influenced by plant physiological reactions. Agric. Eng. 13:199–200.

Kemp, J.G., G.C. Misener, and W.S. Roach. 1972. Development of empirical formulae for drying of hay. Trans. ASAE 15:723–725.

Kepner, R.A., J.R. Goss, J.H. Meyer, and L.G. Jones. 1960. Evaluation of hay conditioning effects. Agric. Eng. 41:299–304.

Klinner, W.E. 1976. A mowing and crop conditioning system for temperate climates. Trans. ASAE 19:237–241.

Klinner, W.E., and G. Shepperson. 1975. The state of hay making technology: A review. J. Brit. Grassl. Soc. 30:259–266.

Koegel, R.G., K.J. Shinners, F.J. Fronczak, and R.J. Straub. 1988. Prototype for production of fast-drying forage mats. Appl. Eng. Agric. 4:126–129.
Krutz, G.W., and D.A. Holt. 1979. For fast drying of forage crops. Agric. Eng. 40:16–17.
Menzies, D.J., and J.R. O'Callaghan. 1971. The effect of temperature on the drying rate of grass. J. Agric. Eng. Res. 16:213–222.
McGechan, M.B. 1990. A cost-benefit study of alternative policies in making grass silage. J. Agric. Eng. Res. 46:153–170.
Morris, R.M. 1972. The rate of water loss from grass samples during hay-type conservation. J. Brit. Grassl. Soc. 27:99–105.
Owen, I.G., and D. Wilman. 1983. Differences between grass species and varieties in rate of drying at 25°C. J. Agric. Sci. (Cambridge) 100:629–636.
Pattey, E., P. Savoie, and P.A. Dube. 1988. The effect of a hay tedder on the field drying rate. Can. Agric. Eng. 30:43–50.
Pedersen, T.T., and W.F. Buchele. 1960a. Drying rate of alfalfa hay. Agric. Eng. 41:86–89, 107, 108.
Pedersen, T.T., and W.F. Buchele. 1960b. Hay-in-a-day harvesting. Agric. Eng. 41:172–175.
Pedro, M.J., and T.J. Gillespie. 1982. Estimating dew duration: II. Utilizing standard weather station data. Agric. Meteorol. 25:297–310.
Person, N.K., and J.W. Sorenson. 1970. Comparative drying rates of selected forage crops. Trans. ASAE 13:352–353, 356.
Peterson, H.B. 1972. Water relationships and irrigation. p. 469–480. *In* C.H. Hanson (ed.) Alfalfa science and technology. Agron. Monogr. 15. ASA, CSSA, and SSSA, Madison, WI.
Pitt, R.E. 1984. Forage drying in relation to pan evaporation. Trans. ASAE 27:1933–1937, 1944.
Pitt, R.E., and M.B. McGechan. 1987. The rewetting of partially dried grass swaths by rain. Dep. Note 1. Scottish Inst. Agric. Eng., Penicuik, Scotland.
Priepke, E.H., and H.D. Bruhn. 1970. Altering physical characteristics of alfalfa to increase the drying rate. Trans. ASAE 13:827–831.
Rijtema, P.E. 1968. On the relation between transpiration, soil physical properties and crop production as a basis for water supply plans. Tech. Bull. 58. Inst. Land and Water Manage. Res., Wageningen, the Netherlands.
Rotz, C.A., S.M. Abrams, and R.J. Davis. 1987. Alfalfa drying, loss and quality as influenced by mechanical and chemical conditioning. Trans. ASAE 30:630–635.
Rotz, C.A., and Y. Chen. 1985. Alfalfa drying model for the field environment. Trans. ASAE 28:1686–1691.
Rotz, C.A., and R.J. Davis. 1986. Sprayer design for chemical conditioning of alfalfa. Trans. ASAE 29:26–30.
Rotz, C.A., and R.J. Davis. 1987. Chemical conditioning of alfalfa with controlled droplet atomizers. Appl. Eng. Agric. 3:275–280.
Rotz, C.A., R.G. Koegel, K.J. Shinners, and R.J. Straub. 1990. Economics of maceration and mat drying of alfalfa on dairy farms. Appl. Eng. Agric. 6:248–256.
Rotz, C.A., and P. Savoie. 1991. Economics of swath manipulation during field curing of alfalfa. Appl. Eng. Agric. 7:316–323.
Rotz, C.A., and D.J. Sprott. 1984. Drying rates, losses and fuel requirements for mowing and conditioning alfalfa. Trans. ASAE 27:715–720.
Rotz, C.A., and J.W. Thomas. 1988. A comparison of chemicals to increase alfalfa drying rate. Appl. Eng. Agric. 4:8–12.
Savoie, P. 1987. Hay tedding losses. Can. Agric. Eng. 30:151–154.
Savoie, P., and S. Beauregard. 1990a. Predicting the effects of hay swath manipulation on field drying. Trans. ASAE 33:1790–1794.
Savoie, P., and S. Beauregard. 1990b. Hay windrow inversion. Appl. Eng. Agric. 6:138–142.
Savoie, P., and A. Mailhot. 1986. Influence of eight factors on the drying rate of timothy hay. Can. Agric. Eng. 28:145–148.
Savoie, P., E. Pattey, and G. Dupuis. 1984. Interactions between grass maturity and swath width during hay drying. Trans. ASAE 27:1679–1683.
Savoie, P., C.A. Rotz, H.F. Bucholtz, and R.C. Brook. 1982. Hay harvesting system losses and drying rates. Trans. ASAE 25:581–585, 589.
Shinners, K.J., R.G. Koegel, and R.J. Straub. 1991. Leaf loss and drying rate of alfalfa as affected by conditioning roll type. Appl. Eng. Agric. 7:46–49.

Sirohi, U.S., R.G. Koegel, and R.J. Straub. 1985. Electrical treatments to increase forage drying rates. Trans. ASAE 28:706–710.

Straub, R.J., and H.D. Bruhn. 1975. Evaluation of roll design in hay conditioning. Trans. ASAE 18:217–220.

Sullivan, J.T. 1973. Drying and storing herbage as hay. p. 1–31. *In* G.W. Butler and R.W. Bailey (ed.) Chemistry and biochemistry of herbage. Vol. 3. Academic Press, London.

van Elderen, E., J. de Feijter, and S.P.J.H. van Hoven. 1972. Moisture in a grass sward. J. Agric. Eng. Res. 17:209–218.

Weeks, S.A., and L.F. Whitney. 1964. Principles of vacuum drying applied to forage. Trans. ASAE 7:452–453.

Whitney, L.F., H.F. Agrawal, and R.B. Livingston. 1969. Stomatal effects on high-temperature, short-time drying of alfalfa leaves. Trans. ASAE 12:769–771.

4 Hay Preservation Effects on Yield and Quality

Michael Collins

University of Kentucky
Lexington, Kentucky

In a survey conducted in 1989 to gauge priorities among >175 manufacturers of farm and specialized equipment, participants were asked to describe and rank constraints to optimum forage production, harvesting, storage, and feeding (Equipment Manufacturers Institute, 1990). Improvement of field drying rates to reduce losses was considered the most important area for future emphasis and the development of harvesting concepts to reduce field losses was next in the forage harvesting area. Within the forage handling and storage area, the development of new concepts for the preservation and storage of forage was considered most critical. Concern about field and storage losses reflects their economic importance to the livestock industry. Waldo (1977) reviewed the literature on chemical preservation of forages and concluded that "much of the potential production of the original crop is lost during harvest and storage."

Losses incurred during the forage harvesting process include mechanical losses as well as respiration and leaching losses (Pitt, 1982). These losses negatively affect both forage quality and yield. Figure 4–1 illustrates field and storage losses for alfalfa (*Medicago sativa* L.) hay stored across a wide range of moistures as hay, low moisture silage, or silage. Physical losses during raking are an important component of the field loss and are affected greatly by hay moisture concentration, especially for legume crops (Fig. 4–2). To avoid large losses, raking should be completed before hay reaches 40% moisture. Grass leaves are more firmly attached to the sheath and stem and thus are less susceptible to loss during raking and baling than are legume leaves (Collins & Sheaffer, 1994).

YIELD AND QUALITY LOSS DURING BALING

Forage species, maturity stage, baler type, and moisture concentration at baling are among the factors that affect hay losses. Forage moisture and package type are two factors that can greatly affect baling losses for alfalfa hay (Table 4–1). Research indicates that baling losses are generally greater for round balers

---

Copyright © 1995 Crop Science Society of Agronomy and American Society of Agronomy, 677 S. Segoe Rd., Madison, WI 53711, USA. *Post-Harvest Physiology and Preservation of Forages.* CSSA Special Publication no. 22.

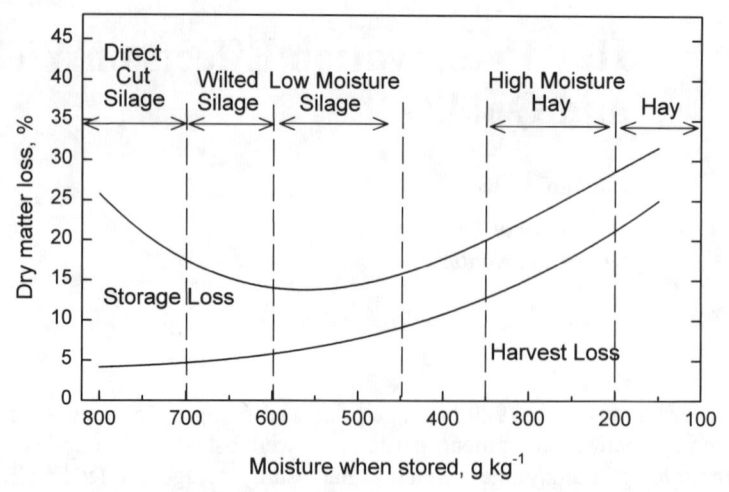

Fig. 4–1. Harvest and storage losses for alfalfa. Harvest losses generally increase with decreasing forage moisture concentration, while storage losses decrease (adapted from Hoglund, 1964).

than for rectangular balers at a given moisture level. Nehrir et al. (1978) reported baling losses as low as 2.6% of the crop dry matter (DM) for alfalfa at 272 g kg^{-1} moisture in small rectangular bales to as much as 13.4% for hay at 150 g kg^{-1} moisture packaged in large stacks. Baling at 245 to 272 g kg^{-1} moisture with a preservative instead of at 150 to 170 g kg^{-1} moisture without treatment reduced baling losses to 6.8% for stacks, but had no significant effect on baling losses for round or rectangular bales. Nighttime baling is sometimes effective in reducing losses further, to an average of 0.8%; however, this practice is not feasible in very humid areas because the hay remoistens quickly at night.

Generally greater mechanical harvesting losses are measured for legume than for grass forage at a given moisture level (McGechan, 1990b). In one field experiment, Klinner (1975) found that even though DM losses during field curing of grass hay were quite high, averaging 19%, alfalfa lost more than twice as

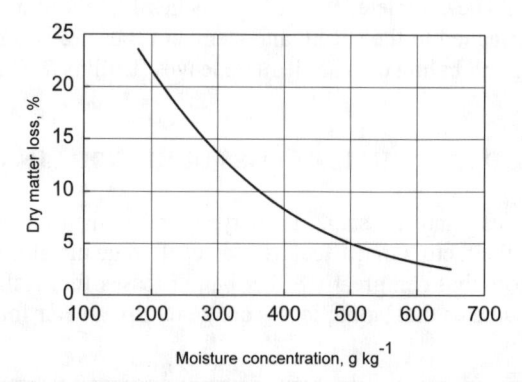

Fig. 4–2. Effects of moisture concentration on raking losses of alfalfa hay. Raking before hay reaches 400 g kg^{-1} moisture concentration avoids excessive losses due to shattering (adapted from Moser, 1980).

Table 4-1. Baling, storage and feeding losses of alfalfa hay baled at 160 or 264 g kg^{-1} moisture and packaged in small rectangular bales, round bales or stacks. Rectangular bales were protected from the weather by storage in a barn while round bales and stacks were stored outside on the ground for 6 months (Nehrir et al., 1978).

Type of losses	Rectangular bale	Round bale	Stack
		% of DM	
160 g kg^{-1} Moisture concentration at baling			
Mowing and curing	13.2	13.2	13.2
Baling	3.5	10.2	13.4
Storage	7.8	11.7	16.8
Feeding	5.2	14.3	16.3
264 g kg^{-1} Moisture concentration at baling			
Mowing and curing	7.1	7.1	7.1
Baling	2.6	9.1	6.8
Storage	4.8	10.1	11.3
Feeding	0.4	5.5	5.5

much, 39%. Cool season grass harvesting and storage losses respond similarly to decreasing moisture, but field losses are less affected by moisture concentration (Parke et al., 1978; McGechan, 1989). Anderson et al. (1981) prepared alfalfa hay windrows from two, three, and four swaths and found that triple windrows reduced baling DM loss to 5 from 14% for the single windrows; however, increased raking losses for triple windrows offset the advantage in reduced baling losses. Shoot morphological differences between legumes and grasses contribute to these differences (Savoie, 1988). More than 90% of the shattered material from alfalfa round bales was found to be leaf (Collins et al., 1987). Leaf tissue

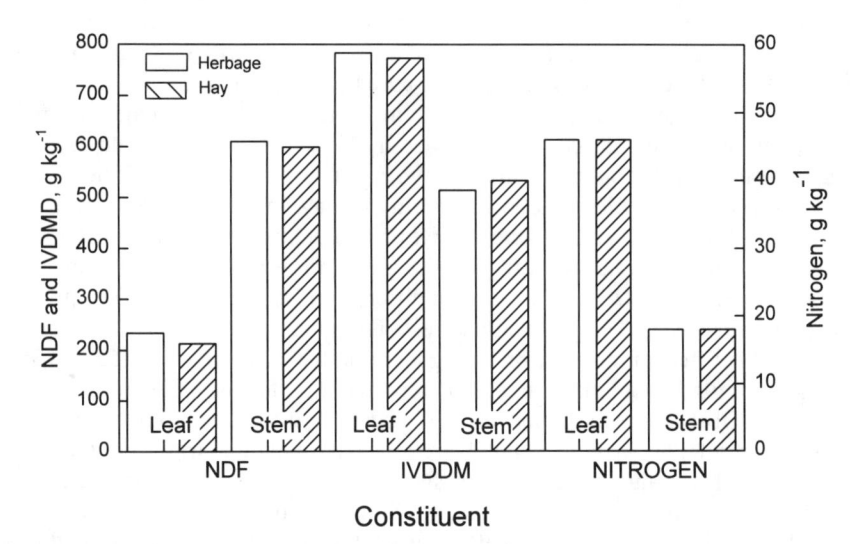

Fig. 4-3. Leaf and stem concentrations of N, neutral detergent fiber (NDF) and in vitro digestible dry matter (IVDMD) in early bloom alfalfa. Leaf tissue is higher in N and IVDMD and lower in NDF than stem tissue in both herbage and hay (Collins, 1991).

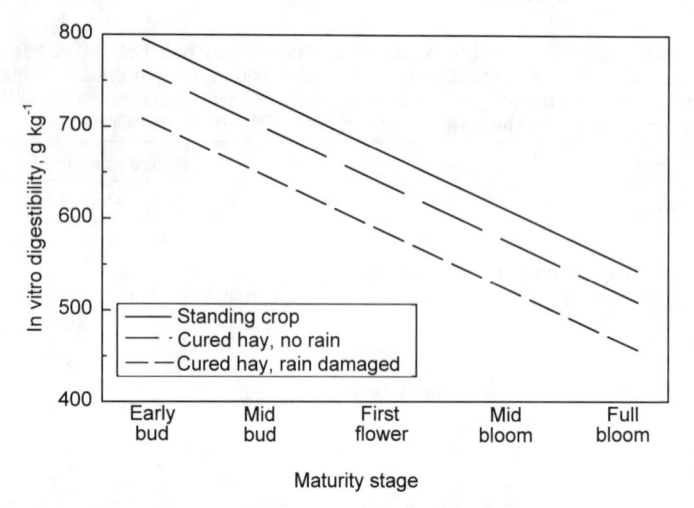

Fig. 4–4. In vitro dry matter digestibility (IVDMD) of alfalfa herbage, cured hay, and rain damaged hay across a range of maturity stages. IVDMD declines with advancing maturity and also is decreased by losses that occur during the hay curing process, especially when rainfall occurs during field exposure (Collins, 1990).

from herbage or hay is much higher in in vitro dry matter disappearance (IVDMD) and N and lower in neutral detergent fiber (NDF) than stem tissue from the same shoots (Fig. 4–3).

Physical shattering of leaf tissue and respiration occurring during field curing reduce hay IVDMD compared with that of the standing herbage (Fig. 4–4). Additional physical loss, respiration, and leaching further reduce quality if rain occurs during the hay curing process. Quality of most forages declines with advancing maturity. The negative impact of advancing maturity on IVDMD of the standing alfalfa crop also is seen in well-cured or rain damaged.

Field research has identified moisture concentration as a major factor in determining leaf loss during baling. Generally, field losses increase with decreasing moisture level at baling (Hundtoft, 1965). Studies frequently indicate a small and somewhat variable reduction in quality of hay baled at very low moisture levels (Collins, 1992). In the latter study, immediately after baling, moist alfalfa hay between 200 and 250 g kg^{-1} was 0.5 g kg^{-1} higher in N, 4 g kg^{-1} lower in acid detergent fiber (ADF), but was not different in NDF and IVDMD compared with field dry hay at 122 g kg^{-1} moisture. With round bales, Nelson et al. (1989a) found little effect of moisture on baling losses in the range between 153 and 357 g kg^{-1}. Similarly, Collins et al. (1987) reported no effect of baling moisture on alfalfa IVDMD in the range between 127 and 377 g kg^{-1} in one trial, but concentrations of total N, ADF, and NDF were affected. In another trial from the same study, DM losses during baling increased from 3.3 to 5.2% as alfalfa dried from 235 to 189 g kg^{-1} moisture. Under good drying conditions, Wittenberg and Moshtaghi-Nia (1990) found no effect of alfalfa moisture between 177 and 299 g kg^{-1}, nor was there any effect on hay concentrations of crude protein (CP), ADF, or NDF.

CHANGES IN YIELD AND QUALITY DURING STORAGE

Compared with silage, storage losses of dry hay are low (McGechan, 1990a); however, storage losses of high moisture hay can be very large (Collins et al., 1987) as can losses of round bales stored under adverse conditions (Lechtenberg et al., 1974). Changes in hay composition during aerobic storage are caused both by respiration and by nonenzymatic chemical reactions (Hlodversson & Kaspersson, 1986). Respiration during storage, measured as CO_2 evolution, was closely related to initial moisture, and was correlated with the increase in hay ADF concentration and the decrease in water soluble carbohydrate concentrations ($r = 0.66$ to 0.85). Dry matter losses, also a reflection of respiration during storage, were more highly correlated with the compositional changes noted above than was CO_2 production.

Microbial Activity During Storage

Except at very low moisture concentrations, plant respiration and microbial activity occur at levels great enough to result in measurable DM losses and composition changes during hay storage. Measurable respiration occurs during hay storage due to plant enzymatic activity even in the absence of visible mold growth (Wood & Parker, 1971) but significant levels of heating are associated with microbial activity. Although numerous bacteria also are present, fungi account for most of the microbial growth observed during hay storage. For example, populations of bacteria in moist grass hay or red clover (*Trifolium pratense* L.) hays changed little over a 3-wk period after baling (Hlodversson & Kaspersson, 1986). Counts of 13 different fungal genera, however, increased sharply early in the storage phase and peaked after ≈ 9 d of storage. These fungi are responsible for most of the detrimental effects that result from storage of moist hay. Previous research reviewed by Rees (1982) indicated that fungal growth in moist hay requires humidities >70% and temperatures >20°C whereas bacteria require humidities of 95%.

Changes in Hay Appearance and Quality During Storage

Hay Color

When moisture concentrations are above ≈ 150 g kg^{-1} at baling, some change in hay color occurs in conjunction with microbial growth and heating during storage. Green color present at baling of moist hay is replaced by various shades of brown. The extent of color change provides a good estimate of the extent of the heating during storage and to the production of Maillard product, which involves a condensation of sugars and amino acids and renders both reactants indigestible (Moser, 1980). At one time, the characteristic brown color of heat damaged hay with the tobacco-like odor was thought to indicate higher quality (Nash, 1985).

Digestibility of Nitrogenous Compounds

Heating effects on protein digestibility can be assessed by determining the change in N concentration of the ADF fraction, referred to as acid detergent in-

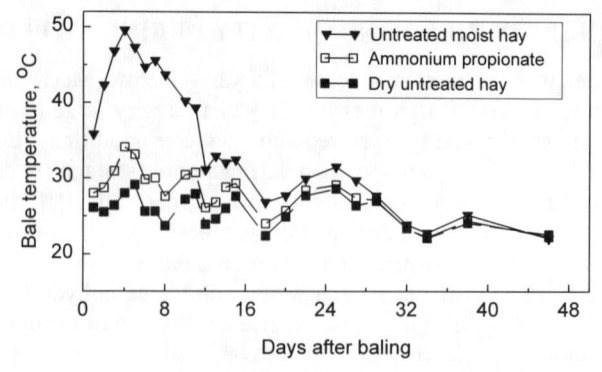

Fig. 4–5. Temperature profiles of stacked alfalfa hay in rectangular bales with 297 or 149 g kg^{-1} moisture. Moist hay treated with ammonium propionate at a rate of 10 kg Mg^{-1} remained cooler during storage than nontreated hay of the same moisture.

soluble nitrogen (ADIN). The microbial activity and heating associated with elevated moisture concentrations at the time of baling also can result in significant reductions in hay quality (Miller et al., 1967). Post-storage concentrations of NDF and ADF were higher for hay with higher initial moisture, but in vivo digestibility by steers (*Bos taurus*) for the same constituents was unaffected. Apparent digestibility coefficients for CP declined from 630 g kg^{-1} for 260 g kg^{-1} moisture alfalfa to as low as 270 g kg^{-1} for hay baled at 585 g kg^{-1} moisture, a response attributed to the browning reaction. The data of Collins et al. (1987) indicated significant bale-type effects on post-storage ADIN levels of alfalfa hay in large round or small rectangular bales. Concentrations of ADIN after storage of alfalfa hay increased linearly with increasing moisture between 150 and 250 g kg^{-1} for both bale types, however, the rate of increase was 1.7-fold greater for round bales.

Dry matter losses generally increase with increased heating during storage of moist hay (Rotz et al., 1988). Kjelgaard et al. (1983) found that losses increased by 1% of the initial hay dry weight for each 10 g kg^{-1} increase in initial moisture >100 g kg^{-1}. Temperatures in stacked rectangular bales of alfalfa at 250 g kg^{-1} moisture peaked near 50°C ≈4 d after baling then declined gradually in temperature over the following week. Figure 4–5. illustrates typical temperature profiles of ammonium propionate-treated and nontreated moist alfalfa at 297 g kg^{-1} compared with field dry hay at 149 g kg^{-1} moisture (Collins, 1992, unpublished data). At moderate moisture levels, peak temperatures are attained within ≈1 wk of storage. In another experiment, native, predominantly grass, hay baled at 340 g kg^{-1} moisture reached similar peak temperatures between 55 and 60°C 1 w after baling (Miller et al., 1967).

Heating

The heat generated by respiration or chemical reactions evaporates some of the water present in the hay at the time of storage and aids in cooling. This heating and evaporation of moisture also contributes to the reduction in hay moisture concentration typically seen during hay storage. Thermal conductivity of dry hay is lower than that of moist hay (Currie & Festenstein, 1971) causing heat transfer to the outside of the stack to become less and less effective as the hay dries. Thus, the internal stack temperatures in hay may increase rapidly after much of the moisture has been removed.

Spontaneous Combustion

Spontaneous combustion becomes possible if microbial heating raises temperatures sufficiently (Miller et al., 1967). Currie and Festenstein (1971) studied heating of timothy (*Phleum pratense* L.), meadow fescue (*Festuca pratense* Hudson), and white clover (*T. repens* L.) hay at 420 to 450 g kg^{-1} moisture. Plant enzymatic activity and microbial growth elevated temperatures to 70°C within a few days when the relative humidity of the air within the hay was maintained at 95 to 97% and within several weeks in drier air. Microbial activity ceased above 70°C and oxidative chemical reactions were responsible for further increases in temperature. The potential for combustion exists if temperatures exceed 170°C.

Currie and Festenstein (1971) calculated that 1.5 m^3 air m^{-3} of hay s^{-1} would be required to heat hay from 20 to 70°C within 30 h. When spontaneous combustion does occur, it appears to do so not in the center of a large stack, but nearer the outside. Oxygen levels in the middle of a large hay stack may be reduced below ambient levels due primarily to microbial activity. Festenstein (1971) suggested that low O$_2$ concentrations beyond a distance of 1 m into the hay stack would limit temperature increases and make spontaneous combustion unlikely. At 165°C hay temperature, the soluble carbohydrates were almost completely eliminated, as were some of the cellulose and hemicellulose (Festenstein, 1971).

Health Effects of Moldy Hay

Mold spores in hay and bedding contribute to a number of respiratory and digestive problems in horses (*Equus caballus*; Naviaux, 1985; Webster et al., 1987; Madelin et al., 1991). Chronic obstructive pulmonary disease, or heaves, is associated with moldy hay and straw. Mold spores also contribute to colic in horses (Naviaux, 1985). Breathing spores of the fungus *Aspergillus fumigatus* during the handling of moldy hay also can cause farmer's lung, a sometimes debilitating disease in which the fungus grows in lung tissue (Lacey & Lord, 1977; Waring & Mullbacher, 1990). Cattle are generally less affected by moldy hay than horses (Lacey & Lord, 1977; Moser, 1980), however, they are subject to mycotic abortion and aspergillosis in response to the presence of certain fungi in moldy hay (Lacey et al., 1978).

Package Size and Storage Method Effects on Hay Yield and Quality

Package Type and Moisture Interactions

Package type, size, and storage method impact storage losses and interact with moisture concentration at baling (Table 4–1; Nehrir et al., 1978). The trend in recent years has been toward larger hay packages that minimize the amount of labor required for baling and transport. A substantial proportion of the U.S. hay crop for on-farm use is stored in large round bales weighing 0.3 Mg or more.

The moisture concentration considered safe for baling hay in small rectangular bales with minimal mold growth is near 200 g kg^{-1}, although lower values are sometimes proposed in dry regions. Higher moisture levels, up to 250 g kg^{-1}, are considered safe for baling nontreated grass hay in England (Harris & Dhanoa, 1984) possibly due to the generally cooler environment. Hay behavior during storage at a particular moisture level also depends upon package type. Recommended maximum moistures given for baling in round bales are usually slightly lower than in rectangular bales, near 180 g kg^{-1}. Russell and Buxton (1985) found that large round bales of mixed alfalfa–orchardgrass (*Dactylis glomerata* L.) hay at 190 g kg^{-1} moisture reached peak temperatures >10°C hotter than similar hay at 156 g kg^{-1} moisture. The application of 1.25 g kg^{-1} sodium diacetate did not affect mean bale temperature. Verma et al. (1985) found that DM loss from ryegrass (*Lolium* sp.) round bales was closely related to baling moisture and averaged 35, 23, and 15% for hay baled at 400, 300, and 200 g kg^{-1} moisture, respectively. Anhydrous ammonia application to plastic covered bales at each moisture level reduced storage losses to 12, 10, and 7%, respectively.

Package Density

Round balers are of two basic designs, one is the fixed chamber baler in which hay accumulates until the mass is sufficient to exert pressure on the forming bale. These balers form packages with lower densities near the center and greater densities near the outside of the package. In variable chamber balers, the bale is formed by belts that exert more uniform pressure during the entire baling process. If both have similar densities in their outer layers, bales with dense cores would be expected to lose a smaller percentage of their DM to weathering because the unweathered core would contain more hay than for bales with low density cores. The data of Verma et al. (1978), who measured average DM losses of 7.7 and 9.1% for alfalfa in uniform density and soft-core round bales, respectively, supports this hypothesis; however, it also has been suggested that restricted heat and moisture transfer from the centers of dense-cored bales might increase DM losses during storage at elevated moisture levels. More mature crops tend to produce lower density round bales (Friesen, 1978). Increasing baling rate by increasing ground speed or by double or triple windrowing reduced the density of alfalfa round bales by as much as 21% (Anderson et al., 1981).

Weathering During Outside Storage

Measured storage losses of round baled hay are generally two or more times losses of similar hay in rectangular bales; however, research indicates that round bales are not inherently subject to greater storage losses than small rectangular bales. They commonly suffer large losses due to adverse storage conditions. Round bales are frequently stored outside without any protection from weathering between baling and feeding. The outer layer of outside stored round bales becomes weathered and contributes greatly to reduced yield and forage quality. Anderson et al. (1981) stored alfalfa round bales inside and measured a DM loss of only 3% compared with 14% for similar bales stored outside over winter in Pennsylvania. Similarly, Belyea et al. (1985) reported DM storage losses for alfalfa round bales ranging from 2.5% for inside bales to 15% for bales stored outside on the

Table 4-2. Fraction of bale volume weathered with different bale sizes and depths of weathered layer.

Bale size (diameter × length)	Depth of weathered layer (cm)				
	5	10	15	20	25
m	Percentage of total bale volume weathered				
1.4 × 1.4	13	27	38	48	58
1.5 × 1.2	11	25	36	46	56
1.7 × 1.7	11	22	32	42	50
1.8 × 1.5	11	21	31	40	48

ground. In Wisconsin, Collins et al. (1987) measured losses of 3.8 and 9.1% after 32 wk for inside and outside stored alfalfa round bales, respectively.

Weathering during outside storage over winter in Pennsylvania affected a layer ≈20 cm deep on the outside of alfalfa round bales of either 1.4 or 1.7 m in diam. (Anderson et al., 1981). Package size affects the proportion of the bale contained in the surface layer and thus the magnitude of losses (Table 4-2). Previous research has indicated that weathering affects primarily the bale circumference rather than the ends (Collins et al., 1987) so these calculations assume uniform losses around the circumference and none on the ends. A weathered layer 15 cm in depth would contain about one-third of the package volume with the proportion decreasing with increasing package size.

Unless other compensating factors exist, larger round bales should have lower weathering losses since a smaller proportion of bale volume would be contained in the surface layer. Research data confirm this (Lechtenberg et al., 1974). The unweathered core of 200-kg round bales contained 78% of the total weight of mixed Kentucky bluegrass (*Poa pratensis* L.), tall fescue (*F. arundinacea* Schreber), and orchardgrass hay in an Indiana study compared with 87% for round bales weighing more than twice as much.

Outside Storage Effects on Hay Quality

Hay chemical composition also is affected by round bale storage method. Table 4-3 shows typical forage quality effects of outside storage of alfalfa round bales (Anderson et al., 1981). The tendency to increase in CP concentration during weathering is typical and also has been observed following rain damage of field-curing hay (Collins, 1983). These data suggest that CP is less subject to loss than the average of other plant constituents under conditions of weathering and inside storage (Rotz et al., 1989). Quality changes during storage differ considerably between the weathered and unweathered portions of round bales. Unweathered hay from round bale cores did not change in IVDMD during storage, but hay from the weathered layer declined sharply to an average IVDMD of only 368 g kg^{-1} after a 5 mo storage period (Lechtenberg et al., 1974).

Protection from Weathering During Outside Storage

Plastic covers placed on individual bales or groups of bales may reduce yield and quality losses during round bale storage (Wood & Parker, 1971; Scales

Table 4-3. Forage quality of the interior and exterior portions of alfalfa round bales stored outside (Anderson et al., 1981).

Bale component	Crude protein	Acid detergent fiber	In vitro dry matter digestibility
		g kg^{-1}	
Interior	189	386	614
Exterior	194	458	469

et al., 1978; Russell & Buxton, 1985). Individual covers reduced the percentage of total bale volume weathered during storage from 23 to 11% (Wood & Parker, 1971). The covers did not improve total DM recovery after storage, but did increase digestible DM recovery and reduced weathering. Under high rainfall conditions in New Zealand, Scales et al. (1978) stored alfalfa round bales with and without plastic covers and found that covers reduced DM loss from 14 to 9%. Stacking alfalfa round bales under plastic reduced losses compared with bales stored on the ground and not covered (Belyea et al., 1985). Russell and Buxton (1985) covered only the upper surface of alfalfa round bales with plastic and also reported reduced weathering losses, and reduced NDF concentration of the weathered layer. Porous mesh bale coverings have been developed for use on round bales with the intention of maintaining bale conformation and providing a smooth surface from which to shed water. Mesh covered alfalfa-smooth bromegrass hay did not differ from twine tied bales in mean IVDMD, NDF, or ADF, but the outer 30 cm layer from mesh covered bales was lower in NDF and higher in IVDMD, indicating less weathering (Russell et al., 1990). Mesh covering also increased recoveries of DM and in vitro digestible DM than twine tied bales.

Avoiding direct contact with the soil surface also may aid in prevention of round bale storage losses. Verma and Nelson (1983) placed ryegrass and alfalfa round bales into storage in May or June and followed changes in yield and quality into the next season. Ryegrass hay baled during May in Louisiana and stored for 48 wk elevated above the soil surface on wooden racks and covered with plastic lost only 11% of their DM during storage; however, ryegrass bales, stored on gravel, on the ground or elevated, but uncovered had losses ranging from 32 to 40%.

Dry matter losses reported for outside round bale storage are generally lower for grass than for legume hay. Under high rainfall conditions, DM losses from alfalfa covered with plastic and elevated above the soil surface, >16 mo., were 45% compared with 11% loss for ryegrass (Verma & Nelson, 1983). The impact of weathering on IVDMD and ADF of grass hay was less than for grass-alfalfa hay in Indiana (Table 4-4; Lechtenberg et al., 1979). Although forage quality was reduced, N concentration increased in the weathered portion, indicating that this constituent alone is not a good indicator of the negative effects of outside storage on hay quality.

Feeding Losses

The weathering that occurs during outside storage of round bales also impacts hay use by livestock. Mature cows consumed 99% of the hay offered for

Table 4-4. Losses of dry matter and quality during storage of round-baled grass and grass-legume hay (Lechtenberg et al., 1979).

Hay type	Fraction	IVDMD†	Total N	NDF†	ADF†	ADL†
			g kg^{-1}			
Grass	Unweathered	588	21.6	699	444	97
	Weathered	425	26.2	625	496	158
Alfalfa-grass	Unweathered	565	22.8	585	450	137
	Weathered	342	27.0	599	487	189

† IVDMD, in vitro dry matter digestibility; NDF, neutral detergent fiber; ADF, acid detergent fiber; ADL, acid detergent fiber.

inside-stored bales, but only 78% of the hay from bales stored outside on the ground (Nelson et al., 1983). Although in vivo DM digestibility of whole bales was not affected by storage method, refusal was reduced from 22% to under 6% by storing outside bales on wooden racks with or without plastic covers.

Lechtenberg et al. (1974) reported that the use of feeding racks to control animal access to round bales reduced wastage during feeding. Thirty-two percent more hay was required to supply the needs of beef cows where hay was fed without racks compared with hay fed in racks. Feeding losses for alfalfa round bales fed to heifers were 12% for inside stored bales, but more than twice that, 25%, for alfalfa stored outside on the ground. Average daily weight gains were reduced to 0.54 kg head^{-1} d^{-1} for heifers consuming round bales stored on the ground compared with 0.65 kg head^{-1} d^{-1} for inside bales and similar gains for heifers fed bales stored outside under plastic. In another study, sheep (*Ovis aries*) consumed 20% more DM of mesh-covered alfalfa round bales than they did of twine-tied bales (Russell et al., 1990). Intake levels of cattle consuming the hay were higher for bales stored on crushed rock than for those stored on the ground.

PRESERVATION OF MOIST HAY

The Basis for Moist Baling

Wilkins (1988) reviewed the literature on forage preservation and concluded that the development of effective treatments to preserve moist hay had great potential as an area for further research. Rain during field curing of hay reduces crop yield and quality due to leaching, leaf loss, and respiration (Rees, 1982; Collins, 1983; Collins, 1985). Extended drying times for legume hay increase the field exposure and thus the likelihood that rain damage will occur. Ignoring weather forecasting, the probability of completing a 4-d hay curing period in Iowa during spring was only 26% (Feyerherm et al., 1966). Reducing the curing period to only 2 d increased the probability of successful field curing to 55%. Pitt (1982) developed a model of forage harvesting and also concluded that shortened field drying times would lead to increased hay yields by reducing harvesting losses. Other factors, such as legume species and grass mixtures also affect drying rates (Collins, 1985).

Equilibrium moisture is the moisture concentration at which hay would stabilize under a given set of uniform temperature, humidity, and radiation con-

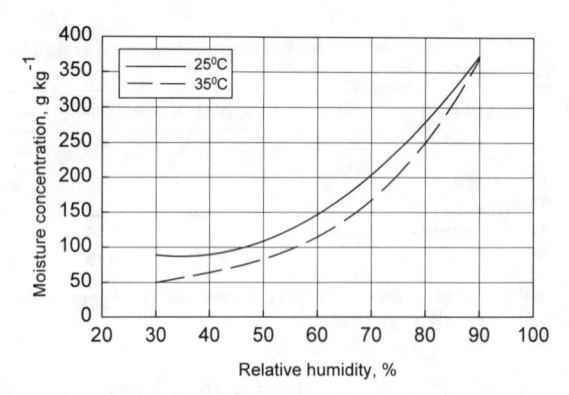

Fig. 4–6. Under a given set of environmental conditions, hay reaches a stable moisture concentration referred to as the equilibrium moisture. Higher relative humidity levels increase equilibrium moisture and higher temperatures reduce equilibrium moisture (adapted from Hill et al., 1977).

ditions. Figure 4–6 illustrates the impact of changing humidity and temperature conditions on equilibrium moisture of alfalfa hay. Under conditions of high relative humidity, the equilibrium moisture concentration may be reached before hay dries sufficiently for safe storage. Depending upon temperature, grass hay would not fall below 200 g kg^{-1} moisture at relative humidities of 78% or more (Nash, 1985). The desire to minimize the duration of field exposure, and to avoid the large physical losses that can occur at low moistures, has led to the development of techniques for the preservation of hay >200 g kg^{-1} moisture.

Types of Compounds and Modes of Action

The addition of compounds or organisms intended to inhibit microbial activity (Tetlow, 1983; Henning et al., 1990), barn drying to remove excess moisture (Parker et al., 1992) and irradiation (Conning, 1983) are among the methods that have been attempted for preservation of moist hay.

A large number of products are available to the forage producer for application to moist hay prior to storage. Among the materials for which beneficial effects on preservation have been noted are sodium diacetate, propionic acid, ammonium propionate, urea, and ammonia (Wood & Parker, 1971; Lacey et al, 1978; Tetlow, 1983). Ammonia-type additives also may improve forage quality by increasing fiber digestibility and adding NPN, especially for grass forages. At the concentrations used in hay preservation these compounds have fungistatic, not fungicidal, activity and thus must be maintained in adequate concentrations throughout the storage period if their benefits are to persist.

Organic Acids

Propionic and other organic acids, at the proper rates, control molding of moist hay by preventing the growth of fungi such as *Aspergillus fumigatus* and *actinomycetes* such as *Micropolyspora faeni* and *Thermoactinomyces vulgaris*, the causative agents in farmer's lung (Lacey et al., 1978).

Woolford (1984) evaluated a number of organic compounds for antimicrobial activity under laboratory conditions. Formic acid, sodium diacetate, propionic acid, ammonium propionate, ammonium dipropionate, and tributyl phosphate were among the compounds tested, each at pH 5, 6, 7, and 8. At pH 5, compared with typical hay pH values near 6, the growth of thermophilic actinomycetes was inhibited by pH alone. Ammonium dipropionate was most effective in inhibition of fungal growth by the propionate compounds followed by ammonium propionate and propionic acid. The salts were more effective against bacteria. Based upon the concentrations used in this controlled study, at pH 5, a rate of ≈2.5 g kg^{-1} of hay weight would be necessary to control fungal and actinomycete growth. This author suggested the use of a mixture of acids and salts to provide the acidifying action that aids in inhibition of fungal growth.

Moist nontreated alfalfa–timothy hay had lower IVDMD and higher ADF and ADIN than hay of similar moisture receiving propionic acid or ammonium isobutyrate at 30 g kg^{-1} or more (Sheaffer & Clark, 1975). Peak storage temperatures of treated hay were 25°C cooler than nontreated moist hay, which peaked near 55°C for nontreated hay. Alfalfa baled at 300 g kg^{-1} moisture, treated with propionic acid at 10 g kg^{-1} hay DM, and stored under plastic maintained lower temperatures than the wet control and similar to that of the field dry control (Khalilian et al., 1990). Storage losses were reduced for alfalfa treated with propionic acid and stored under plastic for 30 d compared with nontreated hay. Hay receiving no treatment or the 5 g kg^{-1} rate of propionic acid had some mold, but hay treated at the higher rate appeared mold-free. Battle et al. (1988) found no improvement in in vivo digestibility by horses of DM, CP, or fiber of alfalfa hay baled at 300 g kg^{-1} moisture and treated with 10 g kg^{-1} fresh weight of 80:20 propionic/acetic acid. Horses consumed hay treated with 10 or 20 g kg^{-1} of 80:20 propionic/acetic acid as well as they did nontreated, field dry control hay.

Organic Acid Application Rates. Inhibition of fungal growth requires maintenance of a minimum acid concentration in the water component of the hay. Thus, higher rates of propionic acid are required to control microbial growth at higher hay moisture concentrations. Under laboratory conditions, hay heating and molding is controlled by propionic acid applications >12.5 g kg^{-1} water, but 30 g kg^{-1} or more may be required under field conditions (Lacey & Lord, 1977; Lacey et al., 1978). Higher than theoretical rates may be needed to overcome variation in hay moisture in the field, losses during application and handling or due to inadequate distribution of the material. In hay with a moisture concentration of 250 g kg^{-1}, a rate of 30 g kg^{-1} of propionic acid would be equivalent to 7.5 kg Mg^{-1} of hay on a fresh weight basis.

Organic Acid Distribution Within the Bale. The effectiveness of hay preservatives depends upon adequate distribution within the package as well as on application of adequate rates. Field variation in moisture concentration can be large and is affected by swath density, initial moisture concentration, and other factors (Collins, 1989). In the latter study, alfalfa bales prepared from a single 1.6 ha field ranged in moisture concentration between 423 and 197 g kg^{-1}. In another field study with alfalfa, individual bale moisture concentrations ranged from 330 to 450 g kg^{-1} when the average moisture was 400 g kg^{-1} (Sheaffer & Clark, 1975).

Inadequate distribution of propionic acid within a bale may allow growth of some fungi (Lacey et al., 1978). These authors concluded that continued growth of certain propionic acid-using microbial species may have reduced the concentration of acid, eventually allowing the growth of other organisms. Nontreated regions within the bale of as little as 3 cm may be sufficient to allow the development of molds within otherwise well preserved hay. Variability in windrow density also causes variation in additive application rate (St. Louis & McCormick, 1988). In most cases, organic acid preservatives have been applied during the baling process by spraying on the incoming windrow, however, Khalilian et al. (1990) developed and tested a device attached to the end of the bale chamber that injected propionic acid after baling.

Duration of Preservation Effects. Results from long-term storage of organic acid-treated hay raise some concerns regarding the duration of preservation action of organic acid-based additives (Rotz et al., 1991). In one study, a gradual temperature increase was observed in alfalfa treated with 15 g kg^{-1} of 80:20 propionic/acetic acid and stored for several months (Van Horn et al., 1988). Similarly, Rotz et al. (1991) found that propionic acid treatment controlled DM losses during 4 wk of storage, but did not affect losses for the same hay 20 wk later. Only ≈15% of the propionic acid applied to mixed grass–clover hay remained after storage periods ranging from 8 to 40 wk (Davies & Warboys, 1978).

Studies of organic acids as hay preservatives frequently indicate positive effects on controlling mold spores (dustiness), but little effect on other storage characteristics such as DM loss or hay color. Hay color was not affected by application of ammonium propionate or a mixture of propionic and acetic acids (Collins, 1992). Rotz et al. (1991) also found that propionic acid treatment at 10 to 20 g kg^{-1} DM on moist alfalfa hay usually reduced visible mold after storage, but failed to improve hay color. Visual observations indicated an increase in red pigmentation on the leaflets of alfalfa treated with either material (Collins, 1992).

Data regarding organic acid treatment of warm season grass hay are not numerous. St. Louis and McCormick (1988) treated moist bermudagrass [*Cynodon dactylon* (L.) Pers.] hay at 323 g kg^{-1} moisture with 8.5 g kg^{-1} of 80:20 propionic/acetic acid before storing the round bales for 28 wk in Mississippi. Crude protein concentrations after storage of acid-treated hay averaged 93 g kg^{-1} and were usually higher than for field dry control hay below 200 g kg^{-1} moisture; however, trends for ADF and IVDMD were less conclusive regarding the effects of acid application. The authors reported little evidence of molding and little treatment effect on that parameter.

Partially Neutralized Acids. In addition to their volatility, organic acids are corrosive to equipment. Partially neutralized organic acids, with pH values near 6, are less volatile and less corrosive than the acid forms while maintaining their effectiveness as preservatives of moist hay. Collins (1992) compared an ammonium propionate product with an 80:20 mixture of propionic and acetic acids on alfalfa near 250 g kg^{-1} moisture. Both products, when applied at 10 g kg^{-1} hay fresh weight, resulted in post-storage dust ratings superior to nontreated wet control bales. Visible mold was nearly absent and no treatment difference in that variable was noted. Post-storage IVDMD of treated hay was equal to the dry control hay and higher than IVDMD of the wet control hay, however, the difference was small.

Adequate rates of sodium diacetate reduced concentrations of NDF and ADIN in round-baled grass hay (Wood & Parker, 1971); however, application at recommended rates may not supply sufficient active ingredient to insure preservation. Russell and Buxton (1985) found that application of 1.25 g sodium diacetate kg^{-1} hay was not sufficient to reduce heating during storage.

Ammonia Compounds

Ammonia has been used successfully in the preservation of moist hays (Moore et al., 1985a,b; Henning et al., 1990) and crop residues (Horton & Steacy, 1979). Ammonia or ammonia-producing compounds can reduce microbial growth during storage of high-moisture hay (Woolford & Tetlow, 1984). Henning et al. (1990) found that urea added at rates of 7 g kg^{-1} DM to tall fescue hay with elevated moisture reduced mold and yeast counts to about one-sixth that of control hay. Woolford and Tetlow (1984) found reduced levels of total viable organisms, molds, thermophilic actinomycetes, and bacteria in grass hay at 400 g kg^{-1} treated with >20 g kg^{-1} ammonia under plastic.

Effects on Fiber Use. The improvement in fiber digestibility following ammonia treatment of high-fiber grass hay may be very large (Moore & Lechtenberg, 1987). Application of 36 g kg^{-1} of either anhydrous ammonia or ammonium hydroxide increased N concentration of orchardgrass hay at 150 or 300 g kg^{-1} moisture. In vitro digestibility of the NDF component, which was present at a concentration of 707 g kg^{-1}, increased from 422 g kg^{-1} for nontreated hay to 682 g kg^{-1} for hay ammoniated at a rate of 36 g kg^{-1} dry weight. Nontreated, moist hay declined >60 g kg^{-1} in organic matter digestibility during 4 wk of aerobic storage whereas ammonia-treated hay increased by up to 120 g kg^{-1} (Woolford & Tetlow, 1984). Henning et al. (1990) found that urea addition to tall fescue hay increased IVDMD and in vitro cell wall disappearance, even for samples collected on the day of baling. Ammoniation of orchardgrass at 30 g kg^{-1} of hay weight under plastic for 7 wk lead to highly significant increases in both average daily gain and in ad libitum DM intake (Moore et al., 1983). Treatment increased DM digestibility from 477 to 548 g kg^{-1} and increased NDF digestibility from 462 to 612 g kg^{-1}.

Moisture Concentration. The beneficial effect of ammoniation on total and ammonium N concentrations is greater at higher hay moistures (Woolford & Tetlow, 1984; Moore et al., 1985a). Ammonia retention after 63 d storage averaged 56, 85, and nearly 100% for moisture concentrations of 100, 300, and 500 g kg^{-1}, respectively. In the same study, NDF digestibility increased from 481 g kg^{-1} for nontreated hay at 100 g kg^{-1} moisture to 568 g kg^{-1} for hay of the same moisture with 12 g kg^{-1} ammonia added. The concentrations of total and nonexchangeable N peaked after ≈4 wk of storage. Woolford and Tetlow (1984) also reported that digestibility of lower moisture hay was affected much less by ammonia treatment than that of high moisture hay.

Table 4-5. Anhydrous ammonia effects on ryegrass quality (Verma et al., 1985).

Moisture concentration/quality constituent	No ammonia	30 g kg^{-1} ammonia
	g kg^{-1}	
400 g kg^{-1} moisture		
Acid detergent fiber	510	444
Crude protein	171	205
In vitro dry matter digestibility	537	651
200 g kg^{-1} moisture		
Acid detergent fiber	403	414
Crude protein	146	155
In vitro dry matter digestibility	537	589

Ryegrass storage losses were reduced by 30 g kg^{-1} of anhydrous ammonia applied to round bales at moistures between 200 and 400 g kg^{-1} (Verma et al., 1985). Hay IVDMD after 20 wk storage was increased from 561 g kg^{-1} for the nontreated control to 651 g kg^{-1} for ammonia treated hay at 400 g kg^{-1} moisture (Table 4–5).

Fiber digestibility of legume hay is affected less by ammonia compounds than that of grass fiber although there may be some response. In an Indiana study (Knapp et al., 1975), tall fescue treated with 10 g kg^{-1} ammonia had 110 g kg^{-1} higher in vitro NDF digestibility, about twice the improvement seen for similar tall fescue–ladino clover (*T. repens* L.). Wittenberg and Moshtaghi-Nia (1990) found no effect on DM or ADF digestibility in vivo of ammoniation at a rate of 32 g kg^{-1} hay DM on alfalfa with 239 g kg^{-1} moisture, although DM retention during storage was improved.

Urea as an Ammonia Source. Urea can also be used to provide ammonia for preservation of hay via the activity of urease (Belanger et al., 1987). Urea provides the ammonia for hay treatment in a form that is easier to handle than anhydrous ammonia. Research suggests that sufficient urease activity is present in moist grass hay to rapidly hydrolyze the added urea to ammonia (Tetlow, 1983). Hay pH of tall fescue at 330 g kg^{-1} moisture, measured just after treatment, increased from 7 to 8 for hay receiving up to 25 g kg^{-1} urea, indicating rapid initiation of urea hydrolysis (Henning et al., 1990). Two days after baling, ≈70% of the urea added to perennial ryegrass hay with a moisture concentration of 430 g kg^{-1} had been hydrolyzed with or without added urease (Tetlow, 1983).

As with ammonia itself, urea addition to moist hay under controlled conditions reduced fungal counts to near one-half peak levels in nontreated hays (Hlodversson & Kaspersson, 1986). Urea treatment increased organic matter digestibility to 588 g kg^{-1} compared with a value of 495 g kg^{-1} in nontreated hay after 8 wk of storage (Tetlow, 1983). An additional 6.6% of the initial DM was retained in the treated hay stored for 16 wk. Urea addition to timothy hay reduced storage temperatures and increased total N concentration after storage (Belanger et al., 1987). Hay at 230 and 290 g kg^{-1} moisture had acceptable control of mold growth during storage when 46 g urea kg^{-1} hay weight. Under the conditions of this study, 35 to 55% of the urea added was actually hydrolyzed. In vitro NDF disappearance was increased 36 g kg^{-1} by the highest urea application rate, a smaller increase than has been observed in some other studies. In another trial with the same species at 230 g kg^{-1} moisture, urea application at 40 g kg^{-1} did not affect in vitro disappearance of either NDF or DM.

Research indicates little advantage, and possibly some disadvantage, to covering urea treated grass hay with plastic (Belanger et al., 1987). About one-half of 46 g kg^{-1} of urea applied to 230 or 290 g kg^{-1} moisture grass hay was retained after a 10 wk storage period. Plastic-covered bales lost less applied urea during storage, but any potential benefit was negated by the absence of drying during the storage period resulting in visible mold development on the exterior surfaces of the stack. Rotz et al. (1990) also found no benefit from covering urea treated alfalfa with plastic, in fact mold development was increased compared with uncovered bales. Over a number of trials, urea between 20 and 60 g kg^{-1} of hay DM improved visual appearance over that of nontreated moist hay, however, field dry hay remained superior in terms of color and mold development.

Cattle consuming hay treated with ammonia compounds sometimes exhibit hyperexcitability that can result in injury or reduced forage intake (Wilkins, 1988). Research indicates that a reaction between ammonia and sugars present in the forage produces a compound 4-methyl imidazole that may be responsible. Ammoniation of lower quality grass hay or crop residues, products for which the NPN and improved fiber use would be greatest, apparently have sufficiently low levels of sugars to avoid the problem.

Microbial Additives

Type of Organisms

The addition of inoculants to enhance silage fermentation has been the topic of considerable research in recent years (Moon & Ely, 1983; Kung et al., 1984; Rust et al., 1989; Van Vurren et al., 1989). Positive responses noted to the addition of *Lactobacillus* and *Pediococcus* sp. to silage include a more rapid pH decline (Kung et al., 1984) and greater stability during feeding. Inoculation in this situation gives the opportunity to add the more efficient homofermentative organisms with the potential to reduce fermentation losses. Kung et al. (1984) found that alfalfa silage between 400 and 700 g kg^{-1} moisture benefited from inoculation with a mixture of *L. plantarum*, *L. brevis*, and *P. acidilactici* by a more rapid initial lactic acid production as some moisture levels and by a lower final pH for the lowest moisture silage.

Behavior in Hay

The possible mode of action of bacterial inoculants in aiding hay preservation under aerobic conditions has been not been clearly elucidated. Wittenberg and Moshtaghi-Nia (1991) treated moist alfalfa with commercial products containing viable lactic acid producing organisms and found no effect on fungal species present in the hay or on plate counts. Average counts of *Aspergillus glaucus*, *Absidia* sp., and *A. flavus* were very low (0.3 × 10^5 organisms per g of DM) on field dry control hay and as high as 45.1 × 10^5 for inoculated alfalfa between 250 and 300 g kg^{-1} moisture. Ammonia-treated alfalfa, included for comparison, had 12.4 × 10⁵ organisms per g of DM. In this study, a commercial product containing nonviable lactic acid bacteria had lower mold counts than the moist control; however, analysis for chitin to quantify fungal biomass showed similar levels in hay inoculated with viable or nonviable organisms. Inoculant-treated and nontreated hays received similar visual ratings for mold development in another study conducted under controlled conditions (Rotz et al., 1988). This study included alfalfa with 200, 250, and 300 g kg⁻¹ moisture, each treated with one of five different microbial inoculant materials, including *L. plantarum* in small plastic bags. Propionic acid treated control hays at each moisture were judged to be mold free. Inoculation at the time of baling had no effect on storage losses. In three of six trials, inoculation increased molding during storage. Collins (1992) applied an aerobic organism, *Bacillus pumilus*, to alfalfa hay between 200 and 220 g kg⁻¹ moisture and found no improvement over the wet control in post-storage visual ratings for mold spores (dustiness), color, or visible mold. Similar alfalfa treated with ammonium propionate showed reduced dustiness levels but was still inferior to the field dry control baled at 120 g kg⁻¹ moisture. Neither inoculant nor ammonium propionate affected moist hay storage losses that averaged 5.6%. Inoculation had no effect on hay IVDMD, NDF, ADF, or ADIN concentrations, but ammonium propionate-treated hay was lower in NDF, ADF, and cellulose.

There are some indications of positive effects of inoculation on hay preservation (Wittenberg & Moshtaghi-Nia, 1990; Tomes et al., 1990), but taken as a whole, they do not exhibit the pattern of changes in hay temperature, composition, and dustiness typical of effective preservatives. One of two products containing viable bacteria, with *L. plantarum* in common, reduced ADIN concentrations for hay at 239 g kg⁻¹ moisture as did a nonviable organism (Wittenberg & Moshtaghi-Nia, 1990); however, hay inoculated with the viable organisms had ADF and NDF concentrations equal to the wet control and showed no improvement in DM digestibility or intake. On the other hand, hay treated with the nonviable organism had higher DM digestibility. Tomes et al. (1990) reported on 19 trials conducted to evaluate a *B. pumilus* inoculant applied to moist alfalfa hay, between 170 and 300 g kg⁻¹ moisture, at the time of baling. After 60 d, fewer bales were visually assessed as very dusty for inoculated hay than for the control bales and color ratings of inoculated bales were higher; however, inoculant treatment had no effect on storage temperatures that averaged 29°C, or concentrations of NDF, ADF, ADFN, and ash. Arledge and Melton (1983) found better visual ratings for mold on alfalfa treated with a mixture of *Lactobacillus* and *Streptococcus* bacteria after 3 to 4 wk of storage. Although the alfalfa in the latter study was relatively low in moisture (197 g kg⁻¹ average for moist hay treatments) 100% of the control bales were rated as having mold and a musty odor.

Nelson et al. (1989a) evaluated a silage inoculant on alfalfa rectangular and round bales containing 434, 265, or 164 g kg⁻¹ moisture and found variable responses depending upon package type and moisture. Heating was reduced by inoculation of 434 g kg⁻¹ moisture rectangular bales whereas inoculated hay at 265 g kg⁻¹ moisture heated more; however, inoculation had no effect on lactic acid concentrations after 39 d of storage and, in contrast to work with rectangular bales, CP concentrations were reduced in round bales (Nelson et al., 1989b).

Barn Drying

Heated or unheated forced air can be used to remove moisture from baled hay prior to storage (Miller, 1946; Parker et al., 1992). Electric fans (610 mm in diam., 3.7–5.2 kW) maintained static pressure near 600 Pa near the duct for alfalfa at 160 kg m^3 DM density (Parker et al., 1992). In 8 of 13 trials drying alfalfa hay from near 350 g kg^{-1} moisture to final moistures of 60 to 120 g kg^{-1}, pre- and post-drying concentrations of CP, NDF, ADF, and IVDMD were not different. In the remaining five trials, changes in composition were small. Solar-heated air 15 to 20°C above ambient temperatures hastened drying significantly. Bales ranging in DM density between 80 and 166 kg m^{-3} were successfully dried except when large variation existed between bales within a batch. Earlier work indicated that the pressure required to force air through bales increased with increasing density and that less pressure was required for bales stacked on edge (Davis & Baker, 1951).

SUMMARY

The extended field curing periods sometimes required to reach safe moistures for hay storage increase the risk of rain damage and may increase physical losses for excessively dry hay. Moist baling and/or hastening drying can shorten field exposure and reduce the risk of rain damage. Moist hay above ≈200 g kg^{-1} moisture is prone to heating due to microbial growth resulting in elevated concentrations of fiber and ADIN after storage and to increased levels of dustiness. Organic acids, including propionic, acetic, and others, and ammonia compounds have generally been shown to control mold growth during storage and to reduce storage temperatures of moist hay. Some positive data are available, but the body of research available for microbial inoculants of moist hay do not demonstrate the pattern of temperature control, post-storage composition and control of dustiness characteristic of effective preservatives. Artificial drying also can be used to successfully remove moisture from baled hay prior to storage.

REFERENCES

Anderson, P.M., W.L. Kjelgaard, L.D. Hoffman, L.L. Wilson, and H.W. Harpster. 1981. Harvesting practices and round bale losses. Trans. ASAE. 24:841–842.

Arledge, J.S., and B. Melton. 1983. Alfalfa hay preservative trial in the Pecos valley. New Mexico State Univ. Agric. Exp. Stn. Res. Rep. 509. Las Cruces.

Battle, G.H., S.G. Jackson, and J.P. Baker. 1988. Acceptability and digestibility of preservative-treated hay by horses. Nutr. Rep. Int. 37:83–89.

Belanger, G., A.M. St. Laurent, C.A. Esau, J.W.G. Nicholson, and R.E. McQueen. 1987. Urea for the preservation of moist hay in big round bales. Can. J. Anim. Sci. 67:1043–1053.

Belyea, R.L., F.A. Martz, and S. Bell. 1985. Storage and feeding losses of large round bales. J. Dairy Sci. 68:3371–3375.

Collins, M. 1983. Wetting and maturity effects on the yield and quality of legume hay. Agron. J. 75:523–527.

Collins, M. 1985. Wetting effects on the yield and quality of legume and legume-grass hays. Agron. J. 77:936–941.

Collins, M. 1989. Conditioning effects and field variation in dry matter concentration of alfalfa hay. p. 1005–1006. *In* Proc. Int. Grassl. Congr. 16th, Nice, France. 4–11 Oct. 1989. Vol. 2. French Grassl. Soc., Versailles Cedex, France.

Collins, M. 1990. Composition and yields of alfalfa fresh forage, field cured hay, and pressed forage. Agron J. 82:91–95.

Collins, M. 1991. Hay curing and water soaking: Effects on composition and digestion of alfalfa leaf and stem components. Crop Sci 31:219–223.

Collins, M. 1992. Chemical, biological and machinery aids for quality haymaking. p. 39–50. *In* Proc. of the Kentucky Alfalfa Conference, 25th, Cave City, KY. 25 Feb. 1992. Kentucky Agric. Exp. Stn., Lexington.

Collins, M., W.H. Paulson, M.F. Finner, N.A. Jorgensen, and C.R. Keuler. 1987. Moisture and storage effects on dry matter and quality losses of alfalfa in round bales. Trans. ASAE 30:913–917.

Collins, M., and C.C. Sheaffer. 1995. Harvesting and storage of cool-season grass hay and silage. *In* L.E. Moser et al. (ed.) Cool season grasses. ASA, Madison, WI (in press).

Conning, D.M. 1983. Evaluation of the irradiation of animal feedstuffs. p. 247–283. *In* P.S. Elias and A.J. Cohen (ed.) Recent advances in food irradiation. Elsevier Biomedical Press, Amsterdam, the Netherlands.

Currie, J.A., and G.N. Festenstein. 1971. Factors defining spontaneous heating and ignition of hay. J. Sci. Food Agric. 2:223–230.

Davies, M.H., and I.B. Warboys. 1978. The effect of propionic acid on the storage losses of hay. J. Br. Grassl. Soc. 33:75–82.

Davis, R.B., Jr., and V.H. Baker. 1951. Fundamentals of drying baled hay. Agric. Eng. 32:21–25.

Equipment Manufacturers Institute. 1990. Hay and forage practices. Priorities for public funded research. Equipment Manufacturers Inst., Chicago, IL.

Festenstein, G.N. 1971. Carbohydrates in hay on self-heating to ignition. J. Sci. Food Agric. 22:231–234.

Feyerherm, A.M., L.D. Bark, and W.C. Burrows. 1966. Probabilities of sequences of wet and dry days in Iowa. Kansas State Univ. Agric. Exp. Stn. Tech. Bull. 139b. Manhattan.

Friesen, O. 1978. Evaluation of hay and forage harvesting methods. p. 317–322. *In* Grain and forage harvesting. Am. Soc. Agric. Eng., St. Joseph, MI.

Harris, C.E., and M.S. Dhanoa. 1984. The drying of component parts of inflorescence-bearing tillers of Italian ryegrass. Grass Forage Sci. 39:271–275.

Henning, J.C., C.T. Dougherty, J. O'Leary, and M. Collins. 1990. Urea for preservation of moist hay. Anim. Feed Sci. Technol. 31:193–204.

Hill, J.D., I.J. Ross, and B.J. Barfield. 1977. The use of vapor pressure deficit to predict drying time for alfalfa hay. Trans ASAE 20:372–374.

Hlodversson, R., and A. Kaspersson. 1986. Nutrient losses during deterioration of hay in relation to changes in biochemical composition and microbial growth. Anim. Feed. Sci. Technol. 15:149–165.

Hoglund, C.R. 1964. Comparative storage losses and feeding values of alfalfa and corn silage crops when harvested at different moisture levels and stored in gas-tight and conventional tower silos: An appraisal of research results. Michigan State Univ, Dep of Agric. Econ. Mimeo 946. East Lansing.

Horton, G.M.J., and G.M. Steacy. 1979. Effect of anhydrous ammonia treatment on the intake and digestibility of cereal straws by steers. J. Anim. Sci. 48:1239–1249.

Hundtoft, E.B. 1965. Handling hay crops: Capacity, quality, losses, power, cost. Cornell Univ. Agric. Eng. Ext. Bull. 363. Ithaca, NY.

Khalilian, A., M.A. Worrell, and D.L. Cross. 1990. A device to inject propionic acid into baled forages. Trans. ASAE 33:36–40.

Kjelgaard, W.L., P.M. Anderson, L.D. Hoftman, L.L. Wilson, and H.W. Harpster. 1983. Round baling from field practices through storage and feeding. p. 657–660. *In* J. A. Smith and V. W. Hays (ed.) Proc. Int. Grassl. Congr., 14th. Lexington, KY. 15–24 June 1981. Westview Press, Boulder, CO.

Klinner, W.E. 1975. Design and performance characteristics of an experimental crop conditioning system for difficult climates. J. Agric. Eng. Res. 20:149–165.

Knapp, W.R., D.A. Holt, and V.L. Lechtenberg. 1975. Hay preservation and quality improvement by anhydrous ammonia treatment. Agron. J. 67:766–769.

Kung, L., Jr., D.B. Grieve, J.W. Thomas, and J.T. Huber. 1984. Added ammonia or microbial inocula for fermentation and nitrogenous compounds of alfalfa ensiled at various percents of dry matter. J. Dairy Sci. 67:299–306.

Lacey, J., and K.A. Lord. 1977. Methods for testing chemical additives to prevent moulding of hay. Annu. Appl. Biol. 87:327–335.

Lacey, J., K.A. Lord, H.G.C. King, and R. Manlove. 1978. Preservation of baled hay with propionic and formic acids and a proprietary additive. Annu. Appl. Biol. 88:65–73.

Lechtenberg, V.L., K.S. Hendrix, D.C. Petritz, and S.D. Parsons. 1979. Compositional changes and losses in large hay bales during outside storage. p. 11–14. *In* Proc. Purdue Cow-Calf Res. Day. West Lafayette, IN. 5 Apr. 1979. Purdue Univ. Agric. Exp. Stn. West Lafayette, IN.

Lechtenberg, V.L., W.H. Smith, S.D. Parsons, and D.C. Petritz. 1974. Storage and feeding of large hay packages for beef cows. J. Anim. Sci. 39:1011–1015.

Madelin, T.M., A.F. Clarke, and T.S. Mair. 1991. Prevalence of serum precipitating antibodies in horses to fungal and thermophilic actinomycete antigens: Effects of environmental challenge. Equine Vet. J. 23:247–252.

McGechan, M.B. 1989. A review of losses arising during conservation of grass forage: 1. Field losses. J. Agric. Eng. Res. 44:1–21.

McGechan, M.B. 1990a. A cost-benefit study of alternative policies in making grass silage. J. Agric. Eng. Res. 46:153–170.

McGechan, M.B. 1990b. A review of losses arising during conservation of grass forage: 2. Storage losses. J. Agric. Eng. Res. 45:1–30.

Miller, L.G., D.C. Clanton, L.F. Nelson, and O.E. Hoehne. 1967. Nutritive value of hay baled at various moisture contents. J. Anim. Sci. 26:1369–1373.

Miller, R.C. 1946. Air flow in drying baled hay with forced ventilation. Agric. Eng. 27:203–208.

Moon, N.J., and L.O. Ely. 1983. Addition of Lactobacillus sp. to aid the fermentation of alfalfa, corn, sorghum, and wheat forages. p. 634–636. *In* J.A. Smith and V.W. Hays (ed.) Proc. Int. Grassl. Congr., 14th. Lexington, KY. 15–24 June 1981. Westview Press, Boulder, CO.

Moore, K.J., and V.L. Lechtenberg. 1987. Chemical composition and digestion in vitro of orchardgrass hay ammoniated by different techniques. Anim. Feed Sci. Technol. 17:109–119.

Moore, K.J., V.L. Lechtenberg, and K.S. Hendrix. 1985a. Quality of orchardgrass hay ammoniated at different rates, moisture concentrations, and treatment durations. Agron. J. 77:67–71.

Moore, K.J., V.L. Lechtenberg, K.S. Hendrix, and J.M. Hertel. 1983. Improving hay quality by ammoniation. p. 626–629. *In* Proc. Int. Grassl. Congr., 14th. Lexington, KY. 15–24 June 1981. Westview Press, Boulder, CO.

Moore, K.J., V.L. Lechtenberg, R.P. Lemenager, J.A. Patterson, and K.S. Hendrix. 1985b. In vitro digestion, chemical composition and fermentation of ammoniated grass and grass-legume silage. Agron. J. 77:758–763.

Moser, L.E. 1980. Quality of forage as affected by post-harvest storage and processing. p. 227–260. *In* C.S. Hoveland (ed.) Crop quality, storage, and utilization. ASA and CSSA, Madison, WI.

Nash, M.J. 1985. Crop conservation and storage in cool temperate climates. Pergamon Press, Oxford.

Naviaux, J.L. 1985. Horses in health and disease. Lea & Febiger, Philadelphia.

Nehrir, H., W.L. Kjelgaard, P.M. Anderson, T.A. Long, L.D. Hoffman, J.B. Washko, L.L. Wilson, and J.P. Mueller. 1978. Chemical additives and hay attributes. Trans. ASAE 21:217–221, 226.

Nelson, M.L., D.M. Headley, and J.A. Loesche. 1989a. Control of fermentation in high-moisture baled alfalfa by inoculation with lactic acid-producing bacteria: II. Small rectangular bales. J. Anim. Sci. 67:1586–1592.

Nelson, M.L., T.J. Klopfenstein, and R.A. Britton. 1989b. Control of fermentation in high-moisture baled alfalfa by inoculation with lactic acid-producing bacteria: I. Large round bales. J. Anim. Sci. 67:1586–1592.

Nelson, B.D., L.I. Verma, and C.R. Montgomery. 1983. Effects of storage method on losses and quality changes in round bales of ryegrass and alfalfa hay. Louisiana Agric. Exp. Stn. Bull. 750. Baton Rouge.

Parke, D., A.G. Dumont, and D.S. Boyce. 1978. A mathematical model to study forage conservation methods. J. Br. Grassl. Soc. 33:261–273.

Parker, B.F., G.M. White, M.R. Lindley, R.S. Gates, M. Collins, S. Lowry, and T.C. Bridges. 1992. Forced-air drying of baled alfalfa hay. Trans. ASAE 35:607–615.

Pitt, R.E. 1982. A probability model for forge harvesting systems. Trans. ASAE 25:549–562.

Rees, D.V.H. 1982. A discussion of sources of dry matter loss during the process of haymaking. J. Agric. Eng. Res. 27:469–479.

Rotz, C.A., D.R. Buckmaster, D.R. Mertens, and J.R. Black. 1989. DAFOSYM: A dairy forage system model for evaluating alternatives in forage conservation. J. Dairy Sci. 72:3050–3063.

Rotz, C.A., R.J. Davis, D.R. Buckmaster, and M.S. Allen. 1991. Preservation of alfalfa hay with propionic acid. Appl. Eng. Agric. 7:33–40.

Rotz, C.A., R.J. Davis, D.R. Buckmaster, and J.W. Thomas. 1988. Bacterial inoculants for preservation of alfalfa hay. J. Prod. Agric. 1:362–367.

Rotz, C.A., J.W. Thomas, R.J. Davis, M.S. Allen, N.L. Schulte Pason, and C.L. Burton. 1990. Preservation of alfalfa hay with urea. Appl. Eng. Agric. 6:679–686.

Russell, J.R., and D.R. Buxton. 1985. Storage of large round bales of hay harvested at different moisture concentrations and treated with sodium diacetate and/or covered with plastic. Anim. Feed Sci. Technol. 13:69–81.

Russell, J.R., S.J. Yoder, and S.J. Marley. 1990. The effects of bale density, type of binding and storage surface on the chemical composition, nutrient recovery and digestibility of large round hay bales. Anim. Feed Sci. Technol. 29:131–145.

Rust, S.R., H.S. Kim, and G.L. Enders. 1989. Effects of a microbial inoculant on fermentation characteristics and nutritional value of corn silage. J. Prod. Agric. 2:235–241.

Savoie, P. 1988. Hay tedding losses. Can. Agric. Eng. 30:39–42.

Scales, G.H., R.A. Moss, and B.F. Quin. 1978. Nutritive value of round hay bales. N.Z. J. Agric. Res. 137:52–53.

Sheaffer, C.C., and N.A. Clark. 1975. Effects of organic preservatives on the quality of aerobically stored high moisture baled hay. Agron. J. 67:660–662.

St. Louis, D.G., and M.E. McCormick. 1988. Hay preservation and storage with propionic:acetic acid and plastic covering. Mississippi Agric. and For. Exp. Stn. Res. Rep. 13(9). Mississippi State.

Tetlow, R.M. 1983. The effect of urea on the preservation and digestibility in vitro of perennial ryegrass. Anim. Feed Sci. and Technol. 10:49–63.

Tomes, N.J., S. Soderlund, J. Lamptey, S. Croak-Brossman, and G. Dana. 1990. Preservation of alfalfa hay by microbial inoculation at baling. J. Prod. Agric. 3:491–497.

Van Horn, H.H., O.C. Ruelke, R.P. Cromwell, M. Ryninks, and K. Oskam. 1988. Effects of chemical drying agents and preservatives on composition and digestibility of alfalfa hay. J. Dairy Sci. 71:2256–2263.

Van Vurren, A.M., K. Bergsma, F. Frol-Kramer, and J.A.C. van Beers. 1989. Effects of addition of cell wall degrading enzymes on the chemical composition and the in sacco degradation of grass silage. Grass Forage Sci. 44:223–230.

Verma, L.R., and B.D. Nelson. 1983. Changes in round bales during storage. Trans. ASAE 26:328–332.

Verma, L.R., B.D. Nelson, and C.R. Montgomery. 1985. Ammonia's effects on high-moisture hay. Stockman (September, 1985):19–21.

Verma, L.R., K. VonBargen, F.G. Owen, and L.J. Perry, Jr. 1978. Characteristics of mechanically formed hay packages after storage. p. 286–289, 299. In G.R. Quick (ed.). Grain and forage harvesting. Proc. Int. Grain and Forage Conf., 1st, Ames, IA. 25–29 Sept. 1977. ASAE, St. Joseph, MI.

Waldo, D.R. 1977. Potential of chemical preservation and improvement of forages. J. Dairy Sci. 60:306–326.

Waring, P., and A. Mullbacher. 1990. Fungal warfare in the medicine chest. New Scientist 27:41–44.

Webster, A.J.F., A.F. Clarke, T.M. Madelin, and C.M. Wathes. 1987. Air hygiene in stables: 1. Effects of stable design, ventilation and management on the concentrations of respirable dust. Equine Vet. J. 19:448–453.

Wilkins, R.J. 1988. The preservation of forages. p. 231–255. In E.R. Orskov (ed.) Feed science. Elsevier Science Publ., New York.

Wittenberg, K.M., and S.A. Moshtaghi-Nia. 1990. Influence of anhydrous ammonia and bacterial preparations on alfalfa forage baled at various moisture levels: I. Nutrient composition and utilization. Anim Feed Sci Technol. 28:333–344.

Wittenberg, K.M., and S.A. Moshtaghi-Nia. 1991. Influence of anhydrous ammonia and bacterial preparations on alfalfa forage baled at various moisture levels: II. Fungal invasion during storage. Anim Feed Sci Technol. 34:67–74.

Wood, J.G.M., and J. Parker. 1971. Respiration during the drying of hay. J. Agric. Eng. Res. 16:179–191.

Woolford, M. 1984. The antimicrobial spectra of organic compounds with respect to their potential as hay preservatives. Grass Forage Sci. 39:75–79.

Woolford, M.K., and R.M. Tetlow. 1984. The effect of anhydrous ammonia and moisture content on the preservation and chemical composition of perennial ryegrass hay. Anim. Feed Sci. Technol. 11:159–166.

5 Legume and Grass Silage Preservation

E.H. Jaster

California Polytechnic State University
San Luis Obispo, California

Silage is a term used to define a product that has undergone fermentation in a silo. More precisely, silage refers to the product of a controlled, anaerobic fermentation of fresh forage in which epiphytic lactic acid bacteria (LAB) convert sugars into lactic acid. Storage structures (silos) establish an anaerobic environment within which fermentation occurs. The success of ensilage is principally dependent upon creation of both sufficient lactic acid bacteria and adequate fermentable carbohydrate in the crop. As a result, the pH decreases and the silage is preserved.

THE ENSILING PROCESS

Various factors have been identified that influence the ensiling process, which include: initial pH, forage buffer capacity, temperature, mass of bacteria, water soluble carbohydrate content, dry matter content, total protein, and total N content, hemicellulose content, and volume of air per volume of herbage (Woolford, 1984; Pitt et al., 1985; Muck, 1988).

The chemical and microbiological characteristics of high quality silage include high lactic acid concentrations relative to concentrations of acetic and butyric acids, low pH, low content of ammonia and volatile N, and low numbers of spore forming anaerobes (Langston et al., 1962a,b; Whittenbury, 1968; McDonald, 1976). Organic acids function as silage preservatives and as energy for ruminants (McDonald et al., 1991). Physically, the criteria used to identify normal silages are green color, pleasant smell and good texture (Newmark et al., 1964).

Respiration

Pitt et al. (1985) has described the ensilage process as having three phases: aerobic, lag, and fermentation. When a forage is harvested, water leaves the plant from surface pores in the surface and cut ends. Surface pores close ≈0.5 to 2 h

---

Copyright © 1995 Crop Science Society of Agronomy and American Society of Agronomy, 677 S. Segoe Rd., Madison, WI 53711, USA. *Post-Harvest Physiology and Preservation of Forages.* CSSA Special Publication no. 22.

after cutting, and drying continues at a much slower rate from the cuticle (McDonald et al., 1991). Plant enzyme processes remain active. Respiration continues until moisture content reaches ≈40% (Noller & Thomas, 1985). Large numbers of aerobic bacteria present on the surface of plant material increase in number as long as O_2 is present.

During the initial phase of ensiling, plant respiration continues within the storage structure as well as other plant enzyme activity such as hydrolysis of cell wall components and proteolysis. Plant respiratory enzymes and aerobic bacteria use available carbohydrates (plant sugar) in crops to produce heat, water, and carbon dioxide (Wylam, 1953; McDonald et al., 1991). The rate and extent of aerobic deterioration depend on an interaction among physical, chemical, and microbiological factors (Ohyama et al., 1975a). The effect of delayed sealing of storage structure is a reduction of carbohydrate supply available both for anaerobic fermentation (lactic acid bacteria) and the animal consuming the silage (Muck, 1988). Reduced lactic acid production may cause silage pH to remain too high to inhibit growth of undesirable microorganisms such as enterobacteria, clostridia, and yeasts (McDonald et al., 1991).

The O_2 trapped in the air spaces within forage in a properly sealed storage structure is consumed rapidly by respiration (Langston et al., 1958). The losses in dry matter (DM) and carbohydrates to remove O_2 trapped in storage structure is minimal (1–2%) (McDonald et al., 1991). Chopped, fresh forage should be well compacted at ensiling to reduce the O_2 available. Maintaining a good seal is especially important to reduce infiltration of O_2 through silo surfaces into silage. Small silo packages, such as bag silos, have a relatively large surface area per mass; therefore, maintaining a good seal is more critical to avoid losses.

Depending on the degree of anaerobic environment, O_2 is depleted from the atmosphere with 90% being removed in 15 min and <0.5% remaining after 30 min (Sprague, 1974; Pitt et al., 1985). Initially, the plant juice liberated from damaged cells only can support a small bacteria population (Greenhill, 1964); however, anaerobic conditions cause cells to rupture, and after a period of time (lag) sufficient juices are liberated to support rapid bacteria growth.

The initial phase lasts longer in bunker silos, especially where packing is poor and more O_2 is trapped. As a result, silage that is stacked or piled typically heats more than other silages. Most high quality silages have had maximum temperatures below 32° C (range from 20 to 30°C), with a mean temperature of 23° C that provides the proper environment for effecting the rate of respiration and growth of lactic acid bacteria (Castle, 1982; Smith et al., 1986). McDonald et al. (1991) suggests that the optimum temperature for LAB is 30° C, while that for clostridia is ≈37° C. Final temperature is dependent on the quantity of air present, initial temperature of forage and air, insulating properties, and specific heat of silage mass (McDonald et al., 1991). Low moisture conditions of crop, improper filling, and incomplete sealing prevent adequate preservation of silage. Leaving chopped forage in a forage wagon for an extended period (8–12 h) delays onset of pH decline, allowing aerobic microbial activity to continue and excessive heating of the silage (Langston et al., 1962a,b; Muck, 1988; Garcia et al., 1989).

Table 5-1. Some lactic acid bacteria of importance during ensiling (McDonald et al., 1991).

Genus	Glucose fermentation	Morphology	Lactate	Species
Lactobacillus	Homofermentative	Rod	DL	*L. acidophilus*
			L(+)	*L. casei*
			DL	*L. curvatus*
			DL	*L. plantarum*
			L(+)	*L. salivarius*
	Heterofermentative	Rod	DL	*L. brevis*
			DL	*L. buchneri*
			DL	*L. fermentum*
			DL	*L. viridescens*
Pediococcus	Homofermentative	Coccus	DL	*P. acidilactici*
			DL	*P. damnosus (cerevisiae)*
			DL	*P. pentosaceus*
Enterococcus	Homofermentative	Coccus	L(+)	*E. faecalis*
			L(+)	*E. faecium*
Lactococcus	Homofermentative	Coccus	L(+)	*L. lactis*
Streptococcus	Homofermentative	Coccus	L(+)	*S. bovis*
Leuconostoc	Heterofermentative	Coccus	D(−)	*L. mesenteroides*

Excess heat may cause increase in Maillard products including acid detergent insoluble N (McDonald et al., 1991). This reduces fermentation potential and lowers nutritive value of the ensiled material.

Fermentation

The fermentation phase objective is to achieve preservation, while minimizing losses of nutrients and avoiding adverse changes in the chemical composition of crop. This requires anaerobic conditions that allow growth of anaerobic organisms, adequate substrate for the lactic acid bacteria in the form of water soluble carbohydrates, low buffering capacity, and a sufficient population of lactic acid bacteria (Muck, 1988; McDonald et al., 1991). Once an anaerobic environment is established, lactic acid bacteria begin to grow rapidly and produce lactic acid alone (homofermentative) as the end product of glycolysis or lactic acid and other products such as acetic acid, mannitol, ethanol, and CO_2 (heterofermentative). The acids reduce silage pH, with lactic acid being most effective. The homofermentative bacteria predominate in good quality silages (Langston et al., 1958). The lactic acid bacteria are found in relatively small numbers on plants (<10 g^{-1}). Release of plant juices, wilting, and chopping with silage harvesting equipment especially in warm weather will increase the number of LAB prior to ensiling (Muck, 1989b). Langston et al. (1958) found populations of LAB ranged from 10^1 to 10^8 g^{-1} herbage. Approximately 10^8 lactic acid bacteria per gram of crop are required before a noticeable drop in pH occurs (Muck, 1988b). Pitt et al. (1985) developed a model of silage fermentation that attempted to predict silage quality. The model indicates initial population and growth rate of LAB affect the rate of pH decline.

A number of species of lactic acid bacteria that are present in silage have been isolated and identified (Table 5–1). Biochemical pathways through which

PHASE 1. Plant material is put into silo.	PHASE 4. Lactic acid formation continues about two more weeks.	PHASE 5. If all has gone properly, silage remains constant.
PHASE 2. Acetic acid is produced.		
PHASE 3. Lactic acid formation begins on third day.		

| 20°C | 32°C | TEMPERATURE CHANGE | 29°C |
| 6.0 | 4.2 | pH CHANGE 4.0 | 3.8 |

ACETIC ACID BACTERIA   LACTIC ACID BACTERIA

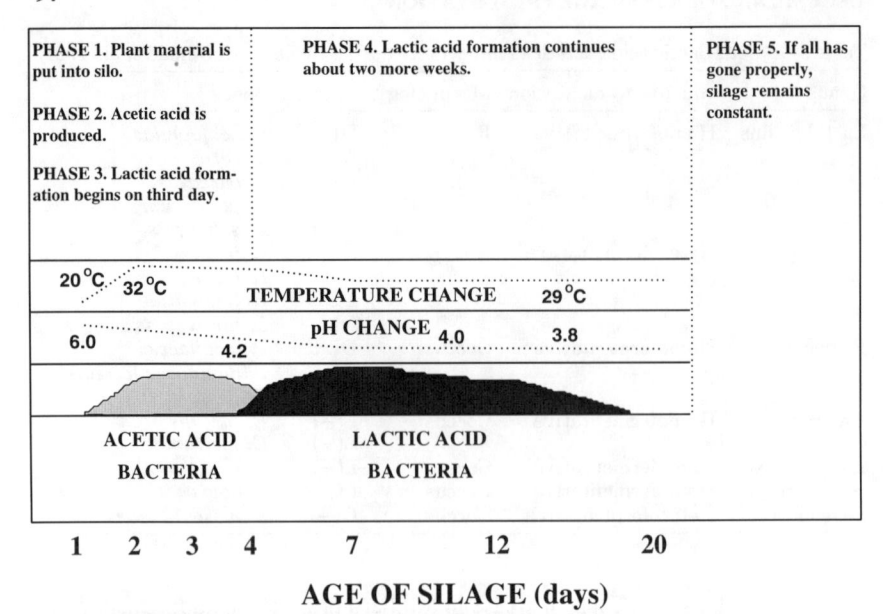

1 2 3 4   7   12   20

AGE OF SILAGE (days)

Fig. 5–1. Five phases of silage fermentation and storage (Kenealy et al., 1982).

LAB metabolize plant sugars have been described (Woolford, 1984; McDonald et al., 1991). Initially, there is a production of small amounts of volatile fatty acids, mainly acetic acid, which is followed by a large amount of lactic acid that preserve the silage (Smith et al., 1986; Fig. 5–1).

Lactic acid bacteria in silage convert readily available carbohydrates to lactic acid, therefore reducing pH of the silage. Relative production of lactic and acetic acid is influenced by the amount of available plant sugar, nutrient species, and pH. During the process of ensiling, lactic acid will represent ≈60% of the total organic acid production, and lactic acid production will peak in 3 to 9 d. Acid production in high quality silage will lower pH to ≈4.2 to 4.0 in high-moisture silage and to 4.5 or lower in wilted silage (Smith et al., 1986). At low pH, bacterial growth ceases, and most enzymatic activity is reduced. This usually occurs within 2 to 3 wk, but may be prolonged with dry (60% DM forage and/or cool temperatures (<10°C). Silage will remain preserved for long periods of time if not exposed to O_2.

Adequate water soluble carbohydrate content is required for growth of LAB and fermentation. Some of the factors influencing water soluble carbohydrate content of plants are: species, cultivar, stage of growth, diurnal variations, climate, and fertilizer level. Legumes and grasses are lower in fermentable carbohydrates and therefore more difficult to ensile than corn (*Zea mays* L.). Legumes are more difficult to ferment than grasses because of higher protein content (Smith et al., 1986) and higher buffering capacity (McDonald et al., 1991). Carbohydrate contents are highest in early spring (high leaf:stem ratio). Monthly variation in carbohydrate content in harvested forage suggests that both light and temperature may influence carbohydrate content. Diurnal variation also occurs, since

Table 5-2. Buffering capacity values of a number of herbage species (McDonald et al., 1991).

Species	Samples	Buffering capacity Range	Mean
		— me kg^{-1} DM —	
Grasses			
Timothy	2	188–342	265
Cocksfoot	5	247–424	335
Italian ryegrass	11	265–589	366
Perennial ryegrass	13	257–558	380
Rhodes grass (*Chloris gayana*)	1	--	435
Legumes			
Red clover	1	--	350
White clover	1	--	512
Lucerne (*Medicago sativa*)	9	390–570	472
Stylo (*Stylosanthes guinanensis*)	1	--	469
Siratro (*Macroptilium atropurpureum*)	1	--	621

carbohydrate content of grasses increases between 0600 and 1800 h, but legumes increase carbohydrate levels between 0600 to 1200 h (McDonald et al., 1991).

Two additional factors that influence successful fermentation and silage pH decline are buffering capacity and dry matter content (Muck, 1988; McDonald et al., 1991). Buffering capacity of plants, or their ability to resist pH change, is an important factor during ensiling. Interest is usually concentrated between pH 6 and 4 since most plant materials have a pH of ≈6, and well-preserved silage is about pH 4. Buffering capacity is expressed as milliequivalent (ME) of acid required to change the pH of 1 kg DM from 6 to 4 (McDonald et al., 1991). In general, legumes are more highly buffered than are grasses (Table 5–2). This conclusion has been confirmed in many studies (McDonald & Henderson, 1962; Muck & Walgenbach, 1985). McDonald and Henderson (1962) have reported the buffering capacity of clovers (*Trifolium* sp.) to be about twice that of grasses, requiring ≈6% lactic acid on a dry weight basis to bring the pH down to 4.0. High buffering capacities of alfalfa have been observed under high K fertilization, with first cutting, and with immature alfalfa. Ensiling these types of alfalfa requires greater concentrations of sugar than later cuttings or mature alfalfa (Muck, 1988).

Dry matter content affects number of bacteria, rate of fermentation, and amount of carbohydrate needed for complete fermentation. Fermentation is restricted as DM content increases. Drier silages tend to stabilize at a higher pH, with lower levels of fermentation acids (Jackson & Forbes, 1970; Leibensperger & Pitt, 1988). For crops below 550 g kg^{-1} DM, a rapid pH decline is essential to maximize quality (Muck, 1988) and to minimize proteolysis (McKersie, 1985).

During ensiling, the amount of acid produced is usually greater than that which could be produced from the fermentation of water soluble carbohydrate alone (Langston et al., 1962a,b; McDonald et al., 1964). The hydrolysis of structural carbohydrates cellulose, hemicellulose, and pectin are suggested to be the main source of these sugars (McDonald et al., 1991). Although some cellulose breaks down during fermentation, the amount is small compared with hemicellulose (McDonald et al., 1962, Morrison, 1979; Pitt et al., 1985). In unwilted forage, 10 to 20% of the hemicellulose is apparently hydrolyzed during ensiling (Morrison, 1979; Moser, 1980). During silage fermentation, hydrolysis of varying amounts of hemicellulose may occur. Lactic acid bacteria can only use soluble sugars as substrate and hemicellulose may be the source of a portion of the lactic and acetic acids produced during ensilage. Hemicellulose is hydrolyzed to 5- or 6-C sugars by chemical hydrolysis or by plant enzymes in forage (Pitt et al., 1985).

In fresh forage, 75 to 90% of the total N is present as protein, the rest being mainly peptides, free amino acids, amides, ureides, nucleotides, chlorophyll, and nitrates (Ohshima & McDonald, 1978). During ensiling, extensive proteolysis results in 40 to 60% of the N being solubilized to nonprotein nitrogenous compounds (peptides, free amino acids, amides, and ammonia; Brady, 1960; Hughes, 1970; Bergen, 1975). Extent of proteolysis decreases with increasing dry matter content of ensilage (Hawkins et al., 1970). Proteolysis of forage will decrease as pH decreases (McDonald et al., 1991). Rapid rates of pH decline are particularly important when ensiling crops of high protein content such as alfalfa (*Medicago sativa* L.), because proteolytic enzyme activity is not inhibited until pH falls to 4.5 to 4.0 (McDonald et al., 1991). Silages of high dry matter and those with excessive air exposure from loose packing are susceptible to excessive heating and the browning reaction with the resultant formation of N containing compounds that are mostly unavailable to ruminants consuming them. Ensiling forages that are too dry increases temperature and the potential for a silo fire from spontaneous combustion (Woolford, 1984).

Critical pH for silage preservation varies with moisture content of crop. Undesirable bacteria, clostridia and enterobacteria, grow well in silages with <30% dry matter (Cranshaw, 1977). Disadvantages of butyric acid production are its weakness as acid (to preserve silage) and large feed energy losses (>20%). Silage energy losses with lactic acid fermentation are low (<5%) (Owens & Prigge, 1975). Evidence exists that spoiled silage from a secondary clostridial fermentation may occur in silage following the primary (i.e., lactic acid) fermentation and is typified by slow rise in silo temperature (cold silage fermentation), high pH, high water soluble N content, high volatile N content, and low contents of lactic acid (McDonald & Edwards, 1976; Muck et al., 1991). Clostridia ferment sugars and lactic acid to butyric acid increasing pH, and some strains degrade amino acids to ammonia (McDonald et al., 1991). Thus, spoiled silage from clostridial fermentation is foul smelling and may depress dry matter intake in ruminants (Smith et al., 1986).

Enterobacteria are nonspore forming, facultative anaerobes that ferment sugars to acetic acid and have the ability to degrade amino acids (Beck, 1978). Both clostridia and enterobacteria are inhibited by low pH. The wetter the silage, the lower the critical pH value. Rapid lactic acid production is important in inhibiting the growth of these undesirable bacteria and reducing fermentation losses. Consequently, factors affecting initial LAB and substrate availability for the LAB have great influence on the development of clostridia and enterobacteria in wet silages (<30% DM). In wilted silage (<35% DM), clostridia are inhibited more by osmotic pressure than by reliance on silage acidity. These silages can have relatively high pH value and low lactic acid content, but little or no butyric acid.

Fungi are eukaryotic heterotrophic microorganisms that grow either as single cells, (yeasts) or as multicellular filamentous colonies (molds). Fungi have a large, if not exclusive, role in the deterioration process in silage made from a variety of forage crops (Woolford, 1984; McDonald et al., 1991). Most fungi need O_2 to grow, although some yeast grow under anaerobic conditions. Yeasts are noted to initiate deterioration or heating of silage on exposure to air. Yeasts are relatively insensitive to pH and known to grow under a pH range of 3 to 8, and some can maintain high populations under anaerobic conditions by fermenting sugars (McDonald et al., 1991; Muck et al., 1991; O'Kiely & Muck, 1992).

Growth of mold and heating in silage is associated with aerobic conditions, such as air leaking into the silage mass, improperly sealed silos, and prolonged wilting of a silage crop prior to ensiling (Ohyama et al., 1975a; Vetter & VonGlan, 1978). Molds are a problem in silage preservation because they break down sugar and lactic acid and hydrolyze cell wall components. In addition, some molds produce substances (mycotoxins) that are harmful to animals and humans (Clark, 1988).

Mathematical models have been developed that predict the inhibitory effect on yeasts by organic acids derived from silage fermentation (Muck et al., 1991). O'Kiely and Muck (1992) found that the inhibitory effect on yeast was not present in herbage, but was present in legumes after fermentation. Management of ensiling and feed out can help reduce aerobic deterioration of silage. The control of deleterious microorganisms by means of effective silage additives would be beneficial. Management of silage making should include a reduction in wilting time to minimize buildup of aerobic microorganisms, rapid silo filling, use of an effective silage additive, and providing and maintaining an adequate seal to the silo.

SILAGE ADDITIVES

Microbial Cultures

Experimental addition of lactic acid bacterial cultures to ensiled forages traces its history to the beginning of this century (Watson & Nash, 1960). In most of the early studies, the reported results were not positive. Many early investigators developed the general principles of silage making. If these principles were followed, a natural population of lactic acid bacteria would ferment the forage, providing adequate amounts of lactic acid for preservation.

The current understanding of the microbiology and fermentation of forage crops has provided significant improvement in our knowledge that growing crops may have low soluble sugar content, may a have a high buffering capacity, and often are poor sources of efficient lactic acid bacteria. (Muck & Speckhard, 1984; Pitt & Leibensperger, 1987; Muck, 1989a; Pitt, 1990).

The value of inoculants for silage has not been definitively clarified. Many studies used small laboratory silos, combined inoculum–fermentation substrate treatments, and used forage crops with varying dry matter and water soluble carbohydrate contents (Anderson et al., 1989). The microbial culture concept involves adding enough LAB to dominate fermentation and reduce the time until rapid lactic acid production begins. Many commercial products claim to improve the rate and extent of silage fermentation, increase bunk life of silage, increase dry matter recovery of silage as well as improve animal performance (Haigh et al., 1987; Hooper et al., 1988). Lactic acid bacteria cultures are primarily marketed as dried or inactive bacteria that become viable when mixed with water or forage. Cultures are marketed in 12 to 22 kg bags for application to forage before ensiling at 0.5 to 1.0 kg per Mg fresh crop. Whittenbury (1961) and McDonald et al. (1991) defined some of the criteria that an organism should satisfy as a potential silage inoculant: (i) it must grow vigorously and be able to compete with and preferably dominate other organisms; (ii) it must be homofermentative to maximize lactic acid production from hexose sugars; (iii) it must be acid tolerant and capable of producing a final pH of at least 4.0 (preferably, it should be able to produce this low pH as rapidly as possible in order to inhibit quickly the activities of other microorganisms); (iv) it must be able to ferment glucose, fructose, sucrose, fructans, and, preferably, pentose sugars; (v) it must not produce dextran from sucrose nor mannitol from fructose; (vi) it should have no action on organic acids; (vii) it should possess a growth temperature range extending to 50° C; and (viii) it should be able to grow in material of low moisture content, as might arise when wilted material is ensiled.

In addition, a lack of proteolytic activity is an essential factor (Woolford, 1984). Factors affecting success of inoculant include type and properties of plants to be ensiled, climatic conditions, ability of inoculated bacteria to grow rapidly in silage, degree of homofermentativeness, and tolerance of low pH (Muck, 1988). The efficiency of lactic acid synthesis from glucose was strain-dependent within the group of homofermentative organisms (McDonald et al., 1991). The inefficiency of many commercial inoculants may be the result of containing LAB species that are not appropriate as silage inoculant or unable to compete effectively with epiphytic flora and/or the use of too low application rates (Pitt, 1990; Nesbakken & Broch-Due, 1991). Whittenbury et al. (1967) and Woolford (1984) have discussed the importance of the composition of an inoculum. Immediately removed as being unsuitable are the *Leuconostoc* sp. (heterofermentative cocci) and heterofermentative lactobacilli, because of their low capacity for producing acid, leaving a selection largely between pediococci and homofermentative lactobacilli (McDonald et al., 1991). McDonald et al., (1991) and Woolford (1984) indicate *Lactobacillus plantarum* has been singled out as a strain that may satisfy the criteria of Whittenbury (1961). Seale and Henderson (1984) ensiled perennial ryegrass (*Lolium perenne* L.), direct cut or wilted, and inoculated with *Lactobacillus plantarum* (10^5 g^{-1}) or a mixture of heterofermentative lactic acid bacteria, *Lactobacillus brevis, lactobacillus buchneri, Leuconostoc dextranicum,* and *Leuconostoc mesenteroides* (10^5 g^{-1}; Table 5–3). The major advantage of the *L. plantarum* over the heterofermentative lactic acid bacteria were lower pH, ammonia-N, acetic acid, and greater lactic acid contents. Gibson et al. (1988) reported that *Lactobacillus plantarum* and *Lactobacillus acidophilus* were the dominant components of homofermentative flora. There is evidence to show that strep-

Table 5-3. Composition of perennial ryegrass as ensiled, and silages treated with and without homofermentative or heterofermentative lactic acid bacteria (10^5 g^{-1}) (Seale & Henderson, 1984; McDonald et al., 1991).

	Grass		Silages					
			Direct cut			Wilted		
	Direct cut	Wilted	Untreated	+Homo	+Hetero	Untreated	+Homo	+Hetero
Dry matter, g kg^{-1}	166	346	--	--	--	--	--	--
pH			4.17	3.84	4.36	4.48	3.82	4.29
Total N, g kg^{-1} DM	28.2	27.0	--	--	--	--	--	--
Ammonia-N, g kg^{-1} DM	--	--	--	--	--	--	--	--
Water soluble carbohydrates, g kg^{-1} DM	157	177	134	70	122	130	50	88
Lactic acid, g kg^{-1} DM	--	--	11	58	9	24	57	24
Acetic acid, g kg^{-1} DM	--	--	132	17	78	89	157	96
Butyric acid, g kg^{-1} DM	--	--	45	16	73	14	9	19
Ethanol, g kg^{-1} DM	--	--	0.0	0.8	0.0	3.3	0.7	0.0
			6.0	0.8	14	11	5	11

tococci and leuconostocs initiate fermentation (pH range 6.5 to 5.0) and are superseded by species of lactobacilli and pediococci as pH falls below 5.0 (Langston et al., 1962a,b; Moon et al., 1981, Fenton, 1987).

To predict silage inoculant effectiveness, one needs to know the level of epiphytic lactic acid bacteria on alfalfa in the forage at harvest. Muck (1989b) counted lactic acid bacteria on alfalfa in the standing crop, at mowing, and after 24, 48, and 72 h of wilting. Few bacteria were found on the standing crop (<10 g^{-1}). Mowing added ≈50 colony forming units (cfu) g^{-1} forage. During wilting the population of LAB increased, with LAB concentrations lowest on top of the swath prior to chopping. Inoculation and growth of microorganisms from farm machinery is aided by the cell solubles being liberated during chopping and laceration. Greenhill (1964) states that the release of plant juice is a prerequisite for the production of significant amounts of lactic acid in good quality silage. Fermentation quality of silage was improved by harvesting with a precision chop (fine chopping) as opposed to a flail harvester (coarse chopping; Apolant & Chestnut, 1985; Gordon, 1982), or by fine chopping versus coarse chopping (Castle et al., 1979). Whittenbury (1968), Rooke (1990) and Rooke et al. (1988) found that populations of LAB at the silage pit ranged from 10^1 to 10^5 g^{-1} of ensiled grass. Pitt and Leibensperger (1987) and Muck (1989a) summarized the literature concerning number of epiphytic LAB found on crops after harvesting by conventional farm machinery and reported levels of 10^3 to 10^7 g^{-1} depending on weather and yield.

Success of an inoculum will be greater if at the time of inoculation, a population is provided that outnumbers and dominates the indigenous population of organisms. Reports indicate additions on the order of 10^6 to 10^7 organisms g^{-1} fresh weight have produced well preserved silages from a variety of forage crops (McDonald et al., 1964; Ohyama et al., 1975a; Carpintero et al., 1979; Ely et al., 1981; Moon et al., 1981; Heron et al., 1988; Bolsen & Hinds, 1984; Kung et al., 1991a,b; Nesbakken & Broche-Due, 1991). Heron et al. (1988) inoculated Italian ryegrass (*L. multiflorum* Lam.) with 10^4, 10^6, or 10^8 organism g^{-1} fresh material. The inoculum was a blend of equal numbers of *Lactobacillus plantarum* and *Pediococcus acidilactici*, with or without addition of glucose (20 g kg^{-1} of fresh material). No beneficial effect of the glucose treatment could be detected. Inoculation with homofermentative bacteria improved silage fermentation and reduced proteolysis. There was no advantage in exceeding 10^6 organisms g^{-1}, but the 10^4 level was insufficient. Most recommendations for inoculant use suggest application rates of at least 10^5, but preferably 10^6 homofermentative lactic acid bacteria g^{-1} fresh crop. Satter et al. (1987) has shown that for responses in animal performance, silage inoculant level must be at least 10 times the epiphytic LAB level on the forage at harvest. Muck (1989a) reported that inoculation at 10% or more of the natural level of lactic acid bacteria on legumes consistently improved the rate of pH decline and shifted fermentation towards lactic acid production. Inoculants applied to forage can reduce final silage pH, increase lactic acid, decrease effluent production, decrease DM loss in silo, and improve performance and milk production of animals fed treated silage (McDonald et al., 1991). Nesbakken and Broch-Due (1991) reported the efficacy of inoculum containing multiple strains of lactic acid bacteria (10^6 g^{-1}) in pilot-scale laboratory silos.

Table 5-4. Daily feed intake and animal performance of grass silage treated with a bacterial inoculant (*Lactobacillus plantarum, Streptococcus faecium,* and *Pediococcus sp.*) or formic acid (Martinsson, 1992).

	Control	Treated Inoculant	Formic Acid
Dry matter, g kg^{-1}	193	194	207
Composition of DM, g kg^{-1}			
Ash	94	95	92
Crude protein	151	157	150
Acetic acid	43	38	18
Propionic acid	5	4	1
Butyric acid	6	1.0	0.4
Lactic acid	64	68	50
Ethanol	6	5	11
pH	4.2	4.1	3.9
Ammonia N, g kg^{-1} total N	96	85	49
Weeks 4–12 Lactation			
Total feed intake, kg DM 100 kg^{-1} LW	3.4	3.4	3.3
Milk yield, kg	23.8	24.7	23.8

Treatment resulted in increased lactic acid levels during initial fermentation, and faster pH drop compared with untreated grasses of low dry matter content. Effect of inoculation on rate of pH fall has been obtained in experiments (Kung et al., 1991b; Anderson et al., 1989; Carpintero et al., 1979). Kung et al. (1991b) investigated the addition of *L. plantarum* (10^6 g^{-1}) on silage fermentation. Addition of inoculant to alfalfa or barley (*Hordeum vulgare* L.) resulted in greater production of lactic acid during ensiling, which caused a more rapid drop in pH during early ensiling. Final chemical composition of silages on Day 60 was not affected by inoculation.

Gordon (1989) reported the inoculation of perennial ryegrass with *Lactobacillus plantarum* (10^6 g^{-1}) and examined the potential of this additive for milk production. Cows (*Bos taurus*) in early lactation fed inoculant treated silage consumed 10% more silage dry matter and produced 2.3 kg d^{-1} more milk than those given the control silages. The effect of adding a mixture of *L. plantarum, Streptococcus faecium,* and *Pediococcus* sp. on the fermentation of timothy (*Phleum pratense* L.) and meadow fescue (*Festuca pratensis* Hudson) ensiled in bunker silos was studied by Martinsson (1992; Table 5–4). Bacterial inoculant was applied at a rate of 1.25×10^5 g^{-1}, and the silage was compared with an untreated silage and one treated with 850 g kg^{-1} formic acid applied at 4 L Mg^{-1}. The silage treated with the inoculant and formic acid were significantly different from control silage in terms of ammonia-N, acetic acid, propionic, and ethanol contents. Cows fed inoculant treated wilted silage produced 4% more milk during early lactation than controls. The authors concluded that higher milk yields from inoculated silage appear to be mediated through increased intake of metabolizable energy.

The bacterial inoculant *Lactobacillus acidophilus* was reported to aid fermentation in some experiments. Moon et al. (1981) added *L. acidophilus* and *Candida* sp., each at 10^4 g^{-1}, to ensiled wheat (*Triticum aestivum* L.), corn, and alfalfa, and obtained a more rapid pH decline and lactic acid concentration in inoculated corn silage; no response was observed in wheat and alfalfa silages. Petit and Flipot (1990) observed no beneficial effect of adding a microbial inoculant mixture on silage composition; however, intake of silage constituents was higher for inoculated than for noninoculated silages, possibly improving animal performance.

Preservation and digestibility were enhanced in wheat silage grown under normal rainfall and environmental temperatures and depressed in drought-stressed wheat forage inoculated with a mixture of *Streptococcus faecium, L. plantarum*, and *Pediococcus acidilacti* (2×10^9 g^{-1}). Microbial-inoculated silage resulted in increased DM and fiber digestibility of wheat silage based rations fed to Holstein heifers (Froetschel et al., 1991); however, inoculants provided no advantage in many research trials. Ely et al. (1982) evaluated the addition of *Lactobacillus acidophilus* and *Candida* sp. (5 g kg^{-1}) to fresh forage in stored concrete stave silos. Data showed no advantage of *L. acidophilus* and *Candida* sp. to crops at ensiling.

Absence of any beneficial effect of inoculation on silage pH may be due to DM content of the silages. Kung et al. (1984, 1987) added inocula to alfalfa wilted to 30, 40, 50, and 60% DM. Microbial additions to alfalfa resulted in increased lactic acid at all DM contents, but final pH was lower than noninoculated silage only at 50 and 60% DM.

Shockey et al. (1985, 1988) evaluated the inoculation of alfalfa with a mixed inoculum (10^4 g^{-1}) of homofermentative lactic acid bacteria. The addition of LAB had no influence on any chemical or microbiological parameter.

The relationship between the effect of inoculation with LAB and addition of water soluble carbohydrate (sugar) to forage is controversial. Ohyama et al. (1973) studied the effect of inoculating forage grass with *Lactobacillus plantarum* (10^6 g^{-1}), with and without addition of glucose (10 g kg^{-1}), but no beneficial effects of inoculation treatment were detected.

In subsequent work, Ohyama et al. (1975b) reported the effect of inoculating Italian ryegrass and cocksfoot (*Dactylis glomerata* L.) with *Lactobacillus plantarum* (10^6 g^{-1}) with or without the addition of glucose (2%). Glucose treatment resulted in large amounts of lactic acid. Changes in pH values and volatile basic N levels confirmed the positive effect of glucose addition and *L. plantarum* inoculation before ensiling.

Seale et al. (1986) compared the effect of sugar and inoculant addition on fermentation of alfalfa silage. They showed that with insufficient sugar in the original crop, bacteria in an inoculant would be unable to produce enough lactic acid to lower pH to an acceptable level.

Jones et al. (1992) ensiled alfalfa treated with sugar (2% fresh weight) and/or with mixed culture of *L. plantarum, S. faecium*, and *Pediococcus acidilactici* (3×10^5 g^{-1} herbage), then examined the fermentation characteristics after 60 d of fermentation (Table 5–5). They indicated that silages were well preserved with inoculation increasing the rate of pH decline for all silage dry matters. Inoculation and sugar addition lowered final pH, acetic acid, ammonia-N, free amino acids, and soluble nonprotein N in silages. The combined treatment also increased lactic acid content with 33 and 43% dry matter silages. The potential nutritional benefit from reducing proteolysis during ensiling requires further investigation.

Table 5-5. Composition of alfalfa silage treated with inoculant and/or sugar (Jones et al., 1992).

	DM	Rate of pH decline	pH	Peptide-N	Ammonia-N	Acetic Acid	Lactic Acid
	g kg^{-1}	d^{-1}		—g kg^{-1} Total N—		g kg^{-1} DM	
Control	330	0.85	4.38	100	64	21.4	89.4
Sugar	330	0.93	4.17	142	55	17.8	104.4
Inoculated	330	2.31	4.22	148	42	11.6	99.5
Inoculated + Sugar	330	1.97	4.05	156	33	8.1	109.5

Reports conducted by other researchers have shown the benefits of including sugars and/or a combination of cell wall degrading enzymes that would increase the fermentation capacity by releasing additional fermentable substrate from cell walls or cell solubles (Woolford, 1984; Herm et al., 1988; Muck, 1988; Kung et al., 1990, 1991b; McDonald, 1991).

Cell Wall Degrading Enzymes

Addition of cellulolytic and hemicellulolytic enzymes as silage additives has been investigated as a method of increasing fermentable sugars (water soluble carbohydrate) and improving the digestibility of organic matter (Leatherwood et al., 1959; Olson & Voelker, 1961; Owen, 1962; McCullough, 1964, 1970; Autrey et al., 1975; Henderson & McDonald, 1977; Buchanan-Smith & Yao, 1981; Henderson et al., 1982; McHan, 1986; Jaster & Moore, 1988).

The process of ensiling is known to effect hydrolysis of structural carbohydrates, especially hemicellulose. Morrison (1979) reported losses of 10 to 20% of the hemicellulose fraction during ensiling of grasses. During the experiment, losses of cellulose were <5%. Henderson and McDonald (1977) applied a cellulase preparation derived from *Aspergillus niger* to ryegrass at a rate of 4 g kg^{-1} of fresh weight. The silage treated with cellulase had increased contents of hydrolyzed cellulose compared with nontreated forage, 361 and 157 g kg^{-1}, respectively. In further investigations, these researchers reported an enzyme preparation from *Trichoderma viride* to be more effective in hydrolyzing cellulose than the similar preparation from *Aspergillus niger*. Owen (1962) applied an enzyme produced by *Aspergillus oryzea* to sorghum [*Sorghum bicolor* (L.) Moench] silage, but it failed to affect the available sugar content of the silages.

Additions of hemicellulase and cellulase mixtures to silage were reported by Jacobs and McAllan (1991; Table 5–6). These authors tested two mixtures of hemicellulases and cellulases (0.4 and 0.2 L Mg^{-1}) ensiled ryegrass. Addition of enzymes reduced levels of cellulose, acid detergent fiber, and neutral detergent fiber compared with those in nontreated silages. Effluent production was highest with enzyme-treated silages. The authors concluded that enzyme additives would be most beneficial on more mature crops of higher DM content. Low levels of water soluble carbohydrate content can be overcome by fermentation of sugars

Table 5-6. The composition of perennial ryegrass silages treated with enzymes (cellulases and hemicellulases; Jacobs & McAllan, 1991).

	Silage Treatment		
	Control	Enzyme 1	Enzyme 2
DM, g kg^{-1}	211	218	217
pH	3.81	3.76	3.80
Composition of DM, g kg^{-1}			
Total N	19.4	19.5	20.2
NH$_3$-N	1.57	1.42	1.52
WSC†	6.07	8.03	6.86
ADF†	346.0	314.0	313.0
NDF†	534.0	513.0	505.0
Cellulose-glucose	277.5	251.2	269.5

† WSC, water soluble carbohydrates; ADF, acid detergent fiber; NDF, neutral detergent fiber.

derived from cell wall polysaccharides, and potential effluent problems may be reduced in drier crops.

Many commercial silage additives contain enzymes with homofermentative lactic acid bacteria (McDonald et al., 1991). Kung et al. (1990) reported the effect of adding either *L. plantarum* and *Pediococcus cerevisiae* (1×10^5 g^{-1}) or a cellulase enzyme complex to barley and vetch (*Vivia sativa* L.) mixture harvested at three stages of maturity. Cellulase activity was 4000 cellulase units g^{-1}. Microbial inoculation reduced silage pH, acetate, and ammonia-N and increased lactic acid concentration when averaged across all maturities. Addition of cellulase enzyme did not improve silage fermentation characteristics.

In a later study, Kung et al. (1991b) reported the effect of adding of cell wall degrading enzymes and *L. plantarum* (1×10^5 g^{-1}) to wilted alfalfa (Table

Table 5-7. Acid detergent fiber (ADF) and neutral detergent fiber (NDF) content of forage treated with microbial inoculant or cellulase and pectinase enzyme complex at 0 and 60 d ensiling (Kung et al., 1991b).

	ADF		NDF	
	0 d	60 d	0 d	60 d
	g kg^{-1} DM			
Effect of inoculant†				
Control	369	362	552	531
Inoculant	354	372	538	536
Effect of Enzyme Complex‡				
0	370	360	558	529
EC-1	366	356	547	547
EC-5	362	378	543	534
EC-50	350	374	534	524

† Microbial inoculant was *Lactobacillus plantarum* and *Pediococcus cerevisiae* added at 1×10^5 g^{-1} forage.
‡ EC = Cellulase and pectinase enzyme complex, EC-1 = A suggested commercial dose of cellulase enzyme (0.6 filter paper units 454 g^{-1} of wet forage and pectinase enzyme (0.02 apple pomface units 454 g^{-1} of wet forage); EC-5 and EC-50 = 5 and 50 times th doses of EC-1, respectively.

5–7). Microbial inoculation improved fermentation, but the cell wall degrading enzyme complex did not affect neutral detergent fiber or acid detergent fiber contents.

Research also has examined the effects of an enzyme mixture and commercial inoculant on silage fermentation, digestibility, and animal performance. Stokes (1992) reported on the effects of adding an enzyme mixture containing cellulase, xylanase, cellobiose, and glucose oxidase (300 mL Mg^{-1}), a commercial multispecies homofermentative LAB culture (176 × 10^9 g^{-1}), or both additives combined to grass–legume forage. Inoculation with and without enzyme mixture, reduced silage pH compared with the control, but inoculation alone was more effective than the combination. Enzyme addition increased dry matter intake and milk production; however, the two silage additives were antagonistic when combined and did not improve silage preservation or animal performance.

McCullough (1970) reported a 10% increase in milk production from haylage with added cellulase without an increase in feed intake. The influence of cellulase was on increased digestion of cellulose. Froetschel et al. (1991) measured the effectiveness of three different microbial inoculant mixtures, and a chemical enzyme silage additive on wheat silage grown under normal and adverse environmental conditions. Preservation and digestibility were enhanced in wheat silage grown under normal rainfall and environmental temperatures and depressed in drought-stressed wheat forage as a result of additive treatment.

Jaster and Moore (1988) reported the effect of an enzyme preparation (0.95 kg Mg^{-1}) having cellulolytic and amylolytic activity on preservation and animal performance of silage produced from bud stage alfalfa. Dry matter losses of haylage were 8.5% for nontreated haylage compared with 4.9% for treated haylage. There were no differences in dry matter intake (DMI) or milk production in lactating cows due to enzyme-treatment.

McHan (1986) studied the effect of adding commercial cellulase to chopped coastal bermudagrass [*Cynodon dactylon* (L.) Pers.] before ensiling in laboratory silos. Cellulase was added to samples at a rate of 10 g kg^{-1} fresh weight, and in vitro DM disappearance was determined. He reported that cellulase-treated silage had a higher water soluble carbohydrate content than the nontreated silage at 30 and 60 d after ensiling. The increase in water soluble content from cellulose treated silage may have resulted from a 35% decrease in cellulose content. Digestibility showed a significant day by treatment effect for 30 and 60 day silage, with the increase due to enzyme treatment less for 60 (4%) than 30 d silage (7%). In experiments conducted by VanVuuren et al. (1989) with grass mixtures, the addition of cellulase reduced cell wall content and pH and increased lactic acid content; however, it had no effect on the digestible organic matter content. Pitt (1990) developed a mathematical model to study the effect of cellulase and amylase additives on rate and extent of fiber digestion and change in fiber concentrations in storage. The model predicts that cellulase addition levels to 5000 g Mg^{-1} silage are required to influence the fermentation process; however, at 100 g Mg^{-1} silage, a significant fraction of cellulase may be hydrolyzed during long storage periods. Additions of amylase at l00 g Mg^{-1} silage is predicted to affect final pH in low water soluble carbohydrate silages.

Review of microbial cultures and cell wall degrading enzymes as silage additives indicates varying degrees of success from the use of such products. Some products reported benefits, while others show no effect. Silage additives are not essential to good silage formation when conditions of moisture and storage are correct. Yet under special circumstances they can be recommended for use. For example, harvesting forage with <30% dry matter, an additive could be beneficial if it encourages a rapid drop in pH and stimulates production of lactic acid (Noller & Thomas, 1985). Otherwise conditions in silage favor production of butyric acid. Most silage additives are not nearly as beneficial if silage contains >30% DM. Additives are not a replacement for a good silo or effective chopping, packing, and sealing practices.

In order to assess the value of a silage additive, Ensminger et al. (1990) recommended that the following criteria be applied: (i) does the product lower the ensiling temperature?; (ii) does the product increase aerobic stability?; (iii) does the product increase dry matter and nutrient recovery from the silo?; (iv) does the product improve feed value and animal performance, particularly when silage is a major ingredient of the ration?; and (v) does the product make for sufficient benefits to offset costs and give a return on investment?

SILAGE MOISTURE CONTENT

Silages may be separated into three groups on basis of moisture level: (i) direct cut (high moisture) silage, 75 to 85% moisture; (ii) wilted silage, 60 to 75% moisture; and (iii) low moisture silage (LMS), 40 to 60% moisture.

Direct-cut silage grasses and legumes are harvested with a forage chopper and immediately stored in silo without any intermediate wilting. Direct cut forage requires a low pH for proper preservation. High moisture content adversely affect fermentation potentially producing a lower quality, unstable silage with a large loss of nutrients due to seepage.

The high moisture level in direct-cut silages can cause nondesirable clostridial growth. Silage produced under these conditions are very aerobically stable, but may have a foul smell, high pH, and reduced intake of DM by ruminants (Noller & Thomas, 1985). Current forage practices recommend some wilting of the forage crop to reduce moisture content prior to ensiling, thus improving preservation and reducing loss of nutrients (Noller & Thomas, 1985; Smith et al., 1986). Also, wilting reduces the weight of crop that is transported from the field to silo and the amount of effluent produced during the ensiling process.

Wilted silage is made from forage which is allowed to dry (wilt) for a short period of time after cutting to reduce moisture content from 60 to 65%. The length of time for a forage to dry is influenced by relative humidity (RH) and plant moisture content. In the field little drying occurs when RH is >60% (Carpentero et al., 1979). Wilting may require only a few hours if good drying conditions exist, or several days under adverse conditions. Crushing or crimping (conditioning) freshly cut forage is normally done to speed up drying. Conditioning breaks the waxy surface of stems and creates more cut ends allowing them to dry at a rate more equal to leaves.

Silage produced by the wilting method still depends on lactic acid produced for preservation; however, there is less fermentation than in direct-cut material. Therefore, a pH of ≈4.5 is typical of a wilted silage (Noller & Thomas, 1985).

Low moisture silages have a reduced moisture (40–60%) content and limited bacterial growth and fermentation. Low moisture silages (also termed haylages) have the advantage of improved DM intake by cattle, reduced fermentation odors, and storage and mechanical feeding equipment available for all silage feeding program. Silages with relatively low moisture are best preserved in sealed, O_2-limiting silos. In some cases, conventional upright silos are used for low moisture haylage, but particular attention must be given to maintaining air-free conditions. The important factors are fine chopping, rapid filling, good sealing, and reduced infiltration of air in silo. Therefore, bunkers, stacks, trench silos, and bag silos are not frequently used at these moisture contents because of the difficulty in maintaining air-tight conditions. Allowing air into haylage will cause heating and the growth of nondesirable yeasts and molds. A disadvantage of low-moisture silage is that it often becomes too dry for good harvesting and storage. Harvest losses increase when the forage is drier, and poor packing and retention of air may result in excessive heat damage of the silage. A report by Goering and Adams (1973) indicated that ≈30% of hay crop silages submitted to state laboratories were heat-damaged.

STORAGE METHODS AND SILOS

Forage characteristics and the type of silo affect silage preservation and storage. The size and type of silo chosen should be influenced by the number and kinds of cattle to be fed, the quantity of the product to be fed, and dry matter losses occurring during storage (Noller & Thomas, 1985). The sources of these losses are initial aerobic losses due to air entrapped in the forage, fermentation losses primarily from the production of CO_2 by anaerobic bacteria, effluent losses, long-term storage losses due to air leaks into the silo and the consequent respiration by plant enzyme or aerobic microorganisms (Pitt, 1986).

There are a wide variety of silos in use: (i) conventional upright (tower) silos (concrete stave, galvanized steel, wood stave, monolithic, tile block, and brick); (ii) gas tight (O_2-limiting) silos (glass lined structures, concrete stave, galvanized steel, and monolithic concrete); (iii) pit silos; (iv) horizontal silos (trench and bunker silos); and (v) temporary silos including enclosed stack silos, open stack silos, modified trench-stack silos, and plastic silos. Estimates of the effect of moisture content of forage to be ensiled on the DM losses in the field and in storage are presented in Table 5–8. Conventional upright (tower) silos are cylindrical in shape and built aboveground. The round shape withstands the pressure of forage against the inner walls. The silo walls need to be smooth and airtight to minimize surface exposure of air to forage. Tower silos are adapted to good packing and should have tight fitting doors. Doors may be sealed with building paper or plastic sheeting to prevent air leaks. Packing and spreading within the silo is effected by a horizontal rotating plate or nozzle. Acid corrosion of

Table 5-8. Estimate of typical dry matter losses in forage stored as silage at different moisture levels based on six months of storage (Shepherd et al., 1953; Moser, 1990).

Silo type/Moisture content of forage as stored	Surface spoilage	Fermentation	Seepage	Total silo losses	Field losses†	From cutting of crop to feeding
			%			
Conventional tower silos						
65 g kg^{-1}	4	8	0	12	4	16
Gas-tight tower silos						
65 g kg^{-1}	0	6	0	6	4	10
50 g kg^{-1}	0	4	0	4	10	14
Trench silos						
85 g kg^{-1}	6	11	10	27	2	29
75 g kg^{-1}	8	9	3	18	2	20
70 g kg^{-1}	10	10	1	21	2	23
Stack silos						
85 g kg^{-1}	12	12	10	34	2	36
75 g kg^{-1}	16	11	3	30	2	32
70 g kg^{-1}	20	12	1	33	2	35

† Losses from forage harvester alone.

walls during fermentation of silages can be reduced with cement resurfacing. The primary advantages of tower silos are durability of structure, minimum top and side spoilage, and convenience of feeding during inclement weather. Upright silos are well adapted to mechanization with mechanical unloaders located at the top or bottom of the silo. Bottom unloaders have the advantage of eliminating silo doors. Tower silos vary in size from 3 to 9 m in diam. and up to 24+ m high. Forage is best stored between 40 and 80% moisture. Relatively high moisture forages (>70%) increase the outward pressure in silo walls and increase the losses of nutrients in effluent (Pitt & Parlange, 1987). Effluent contains high concentrations of water soluble carbohydrates and nitrogenous compounds (McDonald et al., 1960).

Oxygen-limiting silos are tower structures sealed by airtight hatches after filling. Silos operate on a continuous flow principal and have a shell, limiting access of O_2 to silage (Meiering, 1982). The quality of silage in gastight silos is depends on maintaining anaerobic conditions in the head space (Jiang et al., 1989). Advantages include no visible top spoilage, ability to refill at any time, low-moisture material (40–50%) can be ensiled, no silo chute to climb, bottom unloading, and reduced risk of being exposed to lethal silo gas. Some disadvantages are greater cost of construction, slower unloading times than conventional tower silos, and relatively high maintenance costs of O_2-limiting silo unloaders.

Gas exchange between the silo head space in O_2-limiting silos and the environment is a result of the pressure fluctuations in the head space. The dome of sealed silos contains a gas space that forms after settling of the ensiled crop and increases with unloading. Pressure fluctuations are affected by temperature change of the gases in head space or the unloading rate (Meiering, 1986). Breather bags and a pressure relief value are used to reduce the air exchange due to diurnal fluctuations in pressure. Wilted forage is usually stored at 45 to 55% moisture to facilitate unloading from the bottom. On the average, storage losses are lowest in these structures because they are the most air tight; however, initial and annual costs are higher than other types of silos.

Horizontal trench silos are constructed from excavated soil with one end at ground level to permit good drainage and use of machinery. Trench silos are suitable as temporary storage and may be earthen in construction or finished with a concrete floor and side walls.

Bunker silos are constructed aboveground using a concrete floor and wooden or concrete airtight side walls. In comparison to tower silos, bunker silos have the advantages of low initial cost, ease of construction, rapid filling and packing by machinery, particularly for storage of large amounts of forage for large dairy herds. Disadvantages of bunker silos are a greater surface area exposed to air, difficulties in packing and air exclusion (especially when drier forage such as haylage is ensiled), and the inconvenience of feeding in inclement weather. Bunker silos need to be sealed airtight to avoid larger losses from silage spoilage (Buckmaster et al., 1989; Parsons, 1991). Plastic sheeting properly weighted down, (commonly with used tires) is superior to limestone, soil, poor quality roughage, sawdust, or water proof paper as a protective sealer (Gordon, 1967). Plastic covers keep out rain and snow and exclude air from the surface, lowering ensiling temperatures, pH, lactic acid, and nonprotein N concentrations compared with uncovered bunkers (McGuffey & Owens, 1979; Oelberg et al., 1983). Temporary aboveground bunker silos have been constructed using an earthen floor and round bales of hay or straw to form temporary perimeter walls. Greater spoilage of silage would be expected with this system as compared with conventional horizontal bunkers because of greater evaporation and exposure of silage to air. Ideally, spoilage in bunker silos should not exceed the top 10 cm, out of a 4 m deep mass of silage ($\approx$3% spoilage).

Temporary silos include enclosed stack silos, open stack silos, modified trench-stack silos, and plastic silos. In comparison to tower silos, temporary silos have the advantages of low cost, rapid filling and packing, and convenience of location. Stack silos usually are comprised of a pile of forage built vertically aboveground. Surfaces of silage may be left exposed to air or enclosed with straight sides of snow or picket fence, poles or wood staves, and woven wire. The top of stack is either left exposed to air or covered with plastic weighted down by used tires. The amount of spoilage varies from 10 to 50 cm on top and sides of the silage stack. Usually the walls of stack silos are weak and height of the stack should not be greater than twice its diameter. As much as 35% spoilage may occur in stack silos because of the large surface area exposed (Hight & George, 1983); however, with proper packing and sealing silage fermentation losses in stack silos may range from 10 to 14% (Savoie et al. 1986; Savoie, 1988).

Temporary silos constructed of heavy plastic and formed in the shape of a tube have been used successfully in forage feeding programs (Rony et al., 1984). Forage is forced into a plastic sleeve with one end closed, and extension of the sleeve is resisted by a retaining mesh, controlled hydraulically by cable and brake. Plastic silos should be sealed immediately after filling to prevent aeration of silage and large dry matter losses (Henderson & McDonald, 1975). Quality of silage stored in plastic silos is proportional to forage density and the extent of anaerobic environment. Precautions need to be taken to maintain a tight seal, because plastic is subject to tears by machinery, animals, or severe weather. Plastic is removed or cut, then folded back during feedout. It is possible to have cattle self feed silage from plastic bags, but some trampling and wasted feed may result. Plastic is not reusable and may pose a disposal problem. Bags should be located on a well-drained site, preferably paved to avoid problems when unloading in inclement weather. Dry matter losses are close to those found with stave silos, ≈12 to 13% (Noller & Thomas, 1985). Grass silage stored in a plastic silo bag at 42.9% dry matter resulted in total DM losses of 9.0% (Rony et al., 1984).

Round bale silage packaging systems are popular because of their labor efficiency (Nicholson et al. 1991; Fenlon et al., 1989; Harpster et al., 1985). Harpster et al. (1985) outlined the advantages of round bale silage: (i) allows use of hay-making equipment to harvest silage; (ii) does not require silo structures; (iii) can be used to save a mowed field of hay when an anticipated rain storm or extremely high humidity interfere with proper hay curing; (iv) harvesting wilted forage at 50 to 60% moisture reduces leaf loss during baling; since complete field drying is not required, baling time is more predictable; (v) saves about one-third of the harvesting energy and saves fuel compared with silage chopping; and (vi) can be self-fed if properly presented, which saves both labor and fuel.

Round bale silage also has several disadvantages: (i) conditions associated with round bale silages are not optimum for fermentation; (ii) extreme care must be taken to eliminate air leaks and long stems to reduce bale density; (iii) the system requires prompt handling and storage of bales; (iv) machines for lifting and moving heavy, high moisture bales must be available; (v) either individual plastic bags, storage tubes, or plastic sheets to cover group-stacked bales must be purchased; and (vi) plastic is easily damaged and results in forage losses greater than in conventional storages.

Three common methods using plastic materials to produce round bale silage include individual bags, multiple bales in bags, and plastic sheet. Individual bag bale silo systems use various lengths, diameters, and thickness of plastic covering. Bales are lifted with a tractor and spear device and lifted into individual plastic bags. Bags are stored and tied-off in position. Additional plastic can be applied over the top of individual bags. Less labor intensive methods have been developed, including the use of machines that wrap the bale in stretch plastic. Fenlon et al. (1989) found less spoilage (10.2 vs. 21.5% of DM) and lower invisible losses derived from reduction in bale weight during storage (3.3 vs. 6.2% of DM) in wrapped bales than in bagged bales. Nicholson et al. (1991) reported there was a more desirable fermentation pattern in big bales ensiled at 350 to 410 g DM kg^{-1} than those made at 460 to 510 g DM kg^{-1}.

Machinery is available to place several bales in a long plastic tube, which is then sealed at both ends. Producers have found plastic tubes to save labor and be effective for preserving round bale forage; however, more bales will spoil if a bag is torn or opened for long periods during feeding. Round bales also may be stacked under sheets of plastic during storage. Attempts are made to provide airtight seal by covering plastic ends with soil or sand. Problems exist with this system of storage because of the potential for air leaks spoiling a large number of bales.

SUMMARY

This article has attempted to outline the principles of silage fermentation of legume and grass forages as well as providing a practical understanding of silage additives and their use to affect silage fermentation.

While the factors and requirements for fermentation are reasonably well understood, the complex interactions occurring with the addition of microbial inoculants and cell wall degrading enzymes are not well understood. Further, the management and environmental interactions on silage fermentation are profound and far from elucidated.

REFERENCES

Anderson, R., H.I. Gracey, S.J. Kennedy, E.F. Unsworth, and R.W.J. Steen 1989. Evaluation studies in the development of a commercial bacterial inoculant as an additive for grass silage: l. Using pilot-scale tower silos. Grass Forage Sci. 44:361–369.

Apolant, S.M., and D.M. Chestnut. 1985. The effect of mechanical treatment of silage on intake and production of sheep. Anim. Prod. 40:287–296.

Autrey, K.M., T.A. McCaskey, and J.A. Little. 1975. Cellulose digestibility of fibrous materials treated with cellulase. J. Dairy Sci. 58:67–71.

Beck, T. 1978. The microbiology of silage fermentation. p. 61–115. In M.E. McCullough (ed.) Fermentation of silage: A review. Natl. Feed Ingred. Assoc., West Des Moines, IA.

Bergen. W.G. 1975. The influence of silage fermentation on nitrogen utilization. p. 171–180. In Proc. Int. Silage Research Conf.

Bolsen, K.K., and M.A. Hinds. 1984. The role of fermentation aids in silage management. p. 79–112. In M.E. McCullough and K.K. Bolsen (ed.) Silage management. Natl. Feed Ingred. Assoc., West Des Moines, IA.

Brady, C.J. 1960. Redistribution of nitrogen in grass and leguminous fodder plants during wilting and ensilage. J. Sci. Food Agric. ll:276–284.

Buchanan-Smith, J.G., and Y.T. Yao. 1981. Effect of additives containing lactic acid bacteria and/or hydrolytic enzymes with an antioxidant upon the preservation of corn or alfalfa silage. Can. J. Anim. Sci 61:669–680.

Buckmaster, D.R., C.A. Rotz, and R.E. Muck. 1989. A comprehensive model of forage changes in the silo. Trans. ASAE 32:1143–1151.

Carpintero, C.M., A.R. Henderson, and P. McDonald. 1979. The effect of some pretreatments on proteolysis during ensiling of herbage. Grass Forage Sci. 34:311–315.

Castle, M.E. 1982. Making high-quality silage. p. 105–125. In Silage for milk production. Tech. Bull. 2. Natl. Inst. for Res. in Dairying, Reading, England.

Castle, M.E., W.C. Retter, and J.N. Watson. 1979. Silage and milk production: Comparisons between grass silage of three different chop lengths. Grass Forage Sci. 34:293–301.

Clark, A.F. 1988. p. 19–33. In B.A. Stark and J.M. Wilkenson (ed). Silage and health. Chalcombe Publ., Marlow Bottom, England.

Cranshaw, R. 1977. An approach to evaluation of silage additives. ADAS Q. Rev. 24:1–7.

Ely, L.O., N.J. Moon, and E.M. Sudweeks. 1982. Chemical evaluation of *Lactobacillus* addition to alfalfa, corn, sorghum, and wheat forage at ensiling. J. Dairy Sci. 65:1041–1046.

Ely, L.O., E.M. Sudweeks, and N.J. Moon. 1981. Inoculation with *Lactobacillus plantarum* of alfalfa, corn, sorghum, and wheat silages. J. Dairy Sci. 64:2378–2387.

Ensminger, M.E., J.E. Oldfield, and W.W. Heinemann. 1990. Silage/haylage/high moisture grain. p. 331–362. In Feeds and feeding. Ensminger Publ. Company, Clovis, CA.

Fenlon, D.R., J. Wilson, and J.R. Weddell. 1989. The relationship between spoilage and *Listeria monocytogenes* contamination in bagged and wrapped big bale silage. Grass Forage Sci. 44:97–100.

Fenton, M.P. 1987. An investigation into the sources of lactic acid bacteria in grass silage. J. Appl. Bacteriol. 62:181–188.

Froetschel, M.A., L.O. Ely, and H.E. Amos. 1991. Effects of additives and growth environment on preservation and digestibility of wheat silage fed to Holstein heifers. J. Dairy Sci. 74:546–556.

Garcia, A.D., W.G. Olson, D.E. Otterby, J.G. Linn, and W.P. Hansen. 1989. Effects of temperature, moisture, and aeration in fermentation of alfalfa silage. J. Dairy Sci. 72:93–103.

Gibson, T., A.C. Stirling, R.M. Keddie, and R.F. Rosenberger. 1988. Bacteriological changes in silage at controlled temperatures. J. Gen. Microbiol. 19:112–129.

Goering, H.K., and R.S. Adams. 1973. Frequency of heat damaged protein in hay, hay crop silage, and corn silage. J. Anim. Sci. 37:295.

Gordon, C.H. 1967. Storage losses in silage as affected by moisture content and structure. J. Dairy Sci. 50:397–403.

Gordon, F.J. 1982. The effects of degree of chopping grass for silage and method of concentrate allocation on the performance of dairy cows. Grass Forage Sci 37:59–65.

Gordon, F.J. 1989. An evaluation through lactating cattle of a bacterial inoculant is an additive for grass silage. Grass Forage Sci. 44:169–179.

Greenhill, W.L. 1964. Plant juices in relation to silage fermentation: I. The role of the juice. J. Brit. Grassl. Soc. 19:30–37.

Haigh, P.M., M. Appleton, and S.F. Clench. 1987. Effect of commercial inoculant and formic acid ± formalin silage additives on silage fermentation and intake and on live weight change of young cattle. Grass Forage Sci. 42:405–410.

Harpster, H.W., L.L. Wilson, P.M. Anderson, and W.L. Kjelgaard. 1985. New approaches in silage preservation and storage. p. 33–44. *In* Proc. Am. Forage and Grassland Conf., Hershey, PA. 3–6 Mar. 1985. Am. Forage and Grassland Council, Georgetown, TX.

Hawkins, D.R., H.E. Henderson, and D.B. Purser. 1970. Effect of dry matter levels of alfalfa silage on intake and metabolism in the ruminant. J. Anim. Sci 31:617–625.

Henderson, A.R., and P. McDonald. 1975. The effect of delayed sealing on fermentation and losses during ensilage. J. Sci. Food Agric. 26:653–667.

Henderson, A.R., and P. McDonald. 1977. The effect of cellulase preparations on the chemical changes during the ensilage of grass in laboratory silos. J. Sci. Food Agric. 28:468–490.

Henderson, A.R., P. McDonald, and D. Anderson. 1982. The effect of a cellulase preparation derived from *Trichoderma veride* on the chemical changes during the ensilage of grass, lucerne, and clover. J. Sci. Food Agric. 33:16–20.

Herm, S.J.E., R.A. Edwards, and P. McDonald. 1988. The effects of inoculation, addition of glucose and mincing on fermentation and proteolysis in ryegrass ensiled in laboratory silos. Anim. Feed Sci. Technol. 19:85–96.

Heron, S.J.E., R.A. Edwards, and P. McDonald. 1988. The effects of inoculation, addition of glucose and mincing on fermentation and proteolysis in ryegrass ensiled in laboratory silos. Anim. Feed Sci. Technol. 19:85–96.

Hight, W.B., and M.R. George. 1983. Storing silage. Coop. Ext. Leaflet 21332, Univ. California Ext. Service, Davis, CA.

Hooper, P.G., P. Rowlinson, and D.G.Armstrong. 1988. The feeding value of inoculated silage as assessed by use of beef animals. Anim. Prod. 46:526.

Hughes, A.D. 1970. The non-protein nitrogen composition of grass silages: II. The changes occurring during the storage of silage. J. Agric. Sci (Cambridge) 75:421–431.

Jackson, N., and T.J. Forbes. 1970. The voluntary intake by cattle of four silages differing in dry matter content. Anim. Prod. 12:591–599.

Jacobs, J.L., and A.B. McAllan. 1991. Enzymes as silage additives: 1. Silage quality, digestion, digestibility and performance in growing cattle. Grass Forage Sci. 46:63–73.

Jaster, E.H., and K.J. Moore. 1988. Fermentation characteristics and feeding value of enzyme-treated alfalfa haylage. J. Dairy Sci. 71:705–711.

Jiang, S., J.C. Jofriet, and A.G. Meiering. 1989. Breathing of oxygen-limiting tower silos. Trans. ASAE 32:228–231.

Jones, B.A., L.D. Satter, and R.E. Muck. 1992. Influence of bacterial inoculant and substrate addition to lucerne ensiled at different dry matter contents. Grass Forage Sci. 47:19–27.

Kenealy, M.D., M.F. Hutjens, and L.H. Kilmer. 1982. Silage production for dairy cattle. Illinois–Iowa Dairy Guide 205. Univ. of Illinois Coop. Ext. Service, Urbana, IL.

Kung, L., Jr., B.R. Carmean, and R.S. Tung. 1990. Microbial inoculation or cellulase enzyme treatment of barley and vetch silage harvested at three maturities. J. Dairy Sci. 73:1304–1311.

Kung, L., Jr., D.B. Grieve, J.W. Thomas, and J.T. Huber. 1984. Added ammonia on microbial inocula for fermentation and nitorgenous compounds of alfalfa ensiled at various percents of dry matter. J. Dairy Sci. 67:299–306.

Kung, L., Jr., L.D. Satter, B.A. Jones, K.W. Genin, A.L. Sudoma, G.L. Enders, Jr., and H.S. Kim. 1987. Microbial inoculation of low moisture alfalfa silage. J. Dairy Sci. 70:2069–2077.

Kung, L., Jr., R.S. Tung, and K. Maciorowski. 1991a. Effect of microbial inoculant (Ecosyl) and/or glycopeptide antibiotic (vancomycin) on fermentation and aerobic stability of wilted alfalfa silage. Anim. Feed Sci Technol. 35:37–48.

Kung, L., Jr., R.S. Tung, K. G. Maciorowski, K. Buffin, K. Knutsen, and W.R. Aimutis. 1991b. Effects of plant cell-wall degrading enzymes and lactic acid bacteria on silage fermentation and composition. J. Dairy Sci. 74:4284–4296.

Langston, C.W., C. Bouma, and R.M. Conner. 1962a. Chemical and bacteriological changes in grass silage during the early stages of fermentation: II. Bacteriological changes. J. Dairy Sci. 45:618–624.

Langston, C.W., H. Irvin, C.H. Gordon, C. Bouma, H.G. Wiseman, C.G. Melin, and L.A. Moore. 1958. Microbiology and chemistry of grass silage. USDA Tech. Bull. 1187. U.S. Gov. Print. Office, Washington, DC.

Langston, C.W., H.G. Wiseman, C.H. Gordon, W.C. Jacobson, C.G. Melin, L.A. Moore, and J.R. McCalmont. 1962b. Chemical and bacteriological changes in grass silage during the early stages of fermentation: I. Chemical changes. J. Dairy Sci. 45:396–402.

Leatherwood, J.M., R.D. Mochrie, and W.E. Thomas. 1959. Chemical changes produced by a cellulolytic preparation added to silages. J. Anim. Sci. 18:1539.

Leibensperger, R.Y., and R.E. Pitt. 1988. Modeling the effects of formic acid and Molasses on ensilage. J. Dairy Sci. 71:1220–1231.

Martinsson, K. 1992. A study of the efficacy of a bacterial inoculant and formic acid as additives for grass silage in terms of milk production. Grass Forage Sci. 47:189–198.

McCullough, M.E. 1964. Influence of cellulase on silage fermentation. J. Dairy Sci. 47:342.

McCullough, M.E. 1970. Silage research at the Georgia Station. College of Agric. Exp. Stn. Res. Rep. 75. Univ. of Geogia, Athens.

McDonald, P. 1976. Trends in silage making. p. 109–123. *In* I.F.A. Skinner and J.G. Carr (ed.). Microbiology in agriculture, fisheries and food. Academic Press, London.

McDonald, D., and R.A. Edwards. 1976. The influence of conservation methods on digestion and utilization of forages by ruminants. Proc. Nutr. Soc. 35:201–211.

McDonald, P., and A.R. Henderson. 1962. Buffering capacity of herbage samples as a factor in ensilage. J. Sci. Food Agric. 13:395–400.

McDonald, P., A.R. Henderson, and S.J.E. Heron. 1991. The biochemistry of silage. 2nd ed. Chalcombe Publ., Bucks, England.

McDonald, P., A.C. Sterling, A.R. Henderson, W.A. Dewar, G.H. Stark, W.G. Davie, H.T. MacPherson, A.M. Reid, and J. Slater. 1960. Studies on ensilage. Edinburgh School of Agriculture Tech. Bull. 24. Edinburgh, Scotland.

McDonald, P., A.C. Sterling, A.R. Henderson, and R. Whittenbury. 1962. Fermentation studies on wet herbage. J. Sci. Food Agric. 13:581–590.

McDonald, P., A.C. Sterling, A.R. Henderson, and R. Whittenbury. 1964. Fermentation studies on inoculated herbages. J. Sci. Food Agric. 15:429–436.

McGuffey, R.K., and M.J. Owens. 1979. Effects of covering and dry matter at ensiling on preservation of alfalfa in bunker silos. J. Anim. Sci. 49:298–305.

McHan, F. 1986. Cellulase-treated coastal Bermudagrass silage and production of soluble carbohydrates, silage acids, and digestibility. J. Dairy Sci. 69:431–438.

McKersie, B.D. 1985. Effect of pH on proteolysis in ensiled legume forage. Agron. J. 77:81–86.

Meiering, A.G., 1982. Oxygen control in sealed silos. Trans. ASAE 25:1349–1354.

Meiering, A.G., 1986. Pressure compensation for oxygen control in sealed silos. Trans. ASAE. 29:218–222.

Moon, N.J., L.O. Ely, and E.M. Sudweeks. 1981. Fermentation of wheat, corn, and alfalfa silages inoculated with *Lactobacillus acidophilus* and *Candida sp.* at ensiling. J. Dairy Sci. 64:807–813.

Morrison, I.M. 1979. Changes in the cell wall components of laboratory silages and the effect of various additions on these changes. J. Agric. Sci. (Cambridge) 93:581–586.

Moser, L.E. 1980. Quality of forage as affected by post-harvest storage and processing. p. 227–260. *In* C.S. Hoveland (ed.) Crop quality, storage and utilization. CSSA and ASA, Madison, WI.

Muck, R.E. 1988. Factors influencing silage quality and their implications for management. J. Dairy Sci. 71:2992–3002.

Muck, R.E. 1989a. Effect of inoculation on alfalfa silage quality Trans. ASAE 32:1153–1158.

Muck, R.E. 1989b. Initial bacterial numbers on lucerne prior to ensiling. Grass Forage Sci. 44:19–25.

Muck, R.E., R.E. Pitt, and R.Y. Leibensperger. 1991. A model of aerobic fungal growth in silage: I. Microbial chracteristics. Grass Forage Sci. 46:283–299.

Muck, R.E., and M.W. Speckhard. 1984. Moisture level effects on alfalfa silage quality. Am. Soc. of Agric. Eng. Tech. Pap. 84-1532. ASAE, St. Joseph, MI.

Muck, R.E., and R.P. Walgenbach. 1985. Variations in alfalfa buffering capacity. Am. Soc. of Agric. Eng. Tech. Pap. 85-1535. ASAE, St. Joseph, MI.

Nesbakken, T., and M. Broch-Due. 1991. Effects of a commercial inoculant of lactic acid bacteria on the composition of silages made from grasses of low dry matter content. J. Sci. Food Agric. 54:177–190.

Newmark, H., A. Bondi, and R. Volcani. 1964. Amines, aldehydes and ketoacids in silage and their effect on food intake by ruminants. J. Sci. Food Agric. 15:487–492.

Nicholson, J.W.G., R.E. McQueen, E. Charmley, and R.S. Bush. 1991. Forage conservation in round bales or silage bags: effect on ensiling characteristics and animal performance. Can. J. Anim. Sci. 71:1167–1180.

Noller, C.H., and J.W. Thomas. 1985. Hay crop silage. p. 517–527. *In* M.E. Heath et al. (ed.) Forages: The science of grassland agriculture. 4th ed. Iowa State Univ. Press, Ames.

O'Kiely, P., and R.E. Muck. 1992. Aerobic deterioration of lucerne (*Medicago sativa*) and maize (*Zea Maize*) silages-effects of yeasts. J. Sci. Food Agric. 59:139–144.

Oelberg, T.J., A.K. Clark, R.K. McGuffey, and D.J. Schingoethe. 1983. Evaluation of covering dry matter, and preservative at ensiling of alfalfa in bunker silos. J. Dairy Sci. 66:1057–1068.

Ohshima, M., and P. McDonald. 1978. A review of the changes in nitrogenous compounds of herbage during ensilage. J. Sci. Food Agric. 29:497–505.

Ohyama, Y., S. Masaki, and S. Hara. 1975a. Factors influencing deterioration of silages and changes in chemical composition after opening silos. J. Sci. Food Agric. 26:1137–1147.

Ohyama, Y., S. Masaki, and T. Morichi. 1973. Effects of temperature and glucose addition on the process of silage fermentation. Jpn. J. Zootech. Sci. 44:59–66.

Ohyama, Y., T. Morichi, and S. Masaki. 1975b. The effect of inoculation with *Lactobacillus plantarum* and addition of glucose at ensiling on the quality of aerated silages. J. Sci. Food Agric. 26:1001–1008.

Olson, M., and H.H. Voelker. 1961. Effectiveness of enzyme and culture additions on the preservation and feeding value of alfalfa silage. J. Dairy Sci. 44:1204.

Owen, F.G. 1962. Effect of enzymes and bacitracin on silage quality. J. Dairy Sci. 45:934–936.

Owens, F.N., and E.C. Prigge. 1975. Influence of silage fermentation on energy utilization. p. 153–156. *In* Proc. Int. Silage Research Conf, 2nd.

Parsons, D.J. 1991. Modeling gas flow in a silage clamp after opening. J. Agric. Eng. Res. 50:209–218.

Petit, H.V., and P.M. Flipot. 1990. Intake, duodenal flow, and ruminal characteristics of long or short chopped alfalfa-timothy silage with or without inoculant. J. Dairy Sci. 73:3165–3171.

Pitt, R.E. 1986. Dry matter losses due to oxygen infiltration in silos. J. Agric. Eng. Res. 35:193–205.

Pitt, R.E. 1990. A model of cellulase and amylase additives in silage. J. Dairy Sci. 73:1788–1799.

Pitt, R.E. 1990. The probability of inoculant effectiveness in alfalfa silages. Trans ASAE 33:1771–1778.

Pitt, R.E., and R.Y. Leibensperger. 1987. The effectiveness of silage inoculants: A systems approach. Agric. Systems. 25:27–49.

Pitt, R.E., R.E. Muck, and R.Y. Leibensperger. 1985. A quantitative model of the ensilage process in lactate silages. Grass Forage Sci. 40:279–303.

Pitt, R.E., and J.Y. Parlange. 1987. Effluent production from silage with application to tower silos. Trans. ASAE 30:1198–1208.

Rony, D.D., G. Dupuis, and G. Pelletier. 1984. Digestibility by sheep and performance of steers fed silages stored in tower silos and silo press bags. Can. J. Anim. Sci. 64:357–364.

Rooke, J.A. 1990. The numbers of epiphytic bacteria on grass at ensilage on commercial farms. J. Sci. Food Agric. 51:525–533.

Rooke, J.A., F.M. Maya, J.A. Arnold, and D.G. Armstrong. 1988. The chemical composition and nutritive value of grass silages prepared with no additive or with the application of additives containing either *Lactobacillus plantarum* or formic acid. Grass Forage Sci. 43:87–95.

Satter, L.D., J.A. Woodford, B.A. Jones, and R.E. Muck. 1987. Effect of bacterial inoculants on silage quality and animal performance. p. 21–22. *In* Proc. Int. Silage Conf. Sept. 1987. Inst. for Grassland and Anim. Prod., Hurley, England.

Savoie, P. 1988. Optimization of plastic covers for stack silos. J. Agric. Eng. Res. 41:65–73.

Savoie, P., J.M. Fortin, and J.M. Wauthy. 1986. Conservation of grass silage in stack silos and utilization by sheep and dairy cows. Trans. ASAE 29:1784–1789.

Seale, D.R., and A.R. Henderson. 1984. Silage preservation. Proc. of the Silage Conf., 7th, Belfast.

Seale, D.R., A.R. Henderson, K.O. Pettersson, and J.F. Lowe. 1986. The effect of addition of sugar and inoculation with two commercial inoculants on the silage fermentation of lucerne silage in laboratory silos. Grass Forage Sci. 41:61–70.

Shephard, J.B., C.H. Gordon, and L.E. Campbell. 1953. Developments and problems in making grass silage. USDA Bureau Dairy Ind. Inf. Mineo 149.

Shockey, W.L., B.A. Dehority, and H.R. Conrad. 1985. Effects of microbial inoculant on fermentation of alfalfa and corn. J. Dairy Sci 68:3076–3080.

Shockey, W.L., B.A. Dehority, and H.R. Conrad. 1988. Effects of microbial inoculant on fermentation of poor quality alfalfa. J. Dairy Sci. 71:722–726.

Smith, D., R.J. Bula, and R.P. Walgenbach. 1986. Legume and grass silage. p. 231–238. *In* Forage Management. 5th ed. Kendall Hunt Publ. Company, Dubuque, IA.

Sprague, M.A. 1974. Oxygen disappearance in alfalfa silage (*Medicago sativa* L.) p. 651–656. *In* Proc. Int. Grassland Congr. 12th, Moscow.

Stokes, M.R. 1992. Effects of enzyme mixture, an inoculant, and their interaction on silage fermentation and dairy production. J. Dairy Sci. 75:764–773.

VanVuuren, A.M., K. Bergsma, F. Frol-Kramer, and J.A.C. Van Beers. 1989. Effects of addition of cell wall degrading enzymes on chemical composition and the in sacco degradation of grass silage. Grass Forage Sci. 44:223–230.

Vetter, R.L., and K.N. VonGlan. 1978. Abnormal silages and silage related disease problems. p. 281–332. *In* Fermentation of silage: A review. Natl. Feed Ingredient Assoc., West Des Moines, IA.

Watson, S.J., and M.J. Nash. 1960. The conservation of grass and forage crops. Oliver & Boyd, Edinburgh, Scotland.

Whittenbury, R. 1961. An investigation of the lactic acid bacteria. Ph.D. thesis. Univ. of Edinburgh, Scotland.

Whittenbury, R. 1968. Microbiology of grass silage. Process Biochem. 3:27–31.

Whittenbury, R.P. McDonald, and D.G. Bryan-Jones. 1967. A short review of some biochemical and microbiological aspects of ensilage. J. Sci. Food Agric. 18:442–444.

Woolford, M.K. 1984. The chemistry of silage. p. 71–132. *In* The silage fermentation. Marcel Dekker, New York.

Wylam, C.B. 1953. Analytical studies on carbohydrates of grasses and clovers: III. Carbohydrate breakdown during wilting and ensilage. J. Sci. Food Agric. 4:527–531.